TYRES
OVER A CUP OF COFFEE

RAJNISH AGARWAL

This book **"Tyres Over A Cup of Coffee"** by Rajnish Agarwal is self-published by the author.

All the printing processes of the book is powered by

24by7 Publishing

13 New Road, Kolkata - 51, India

https://www.24by7publishing.com

mail@24by7publishing.com

+91 9831 470 133

+91 9433 444 334

Cover Design by Thausif Ahamed

MRP: INR 990.00

First Published in December 2021

Version 1.00

ISBN: 978-93-5578-138-3

DECLARATION

All the royalties received on this book, net off expenses and taxes, if any, shall be donated to 'UNICEF- India'.

DEDICATED TO

All auto enthusiasts and vehicle owners who want to take better and smart care of their vehicle's tyres.

My daughter, Palak, for being my strength and the editor of this compilation.

DISCLAIMER / DECLARATION

This is to declare that the write-ups are mine and I being associated with JK, Michelin and Bridgestone Tyres franchise concept stores, have used my experience of over 2 decades to pen down all information that I have.

All the data in figures has been sourced as mentioned and some self-made, after analysis which would show approximate figures and may not be cent percent accurate. Other data and figures sourced from different websites are mentioned therein. The aim is to project an idea to the readers of the tyre industry's performance in India and globally.

The book is not sponsored, endorsed or otherwise affiliated with any Indian or Multi-National tyre company.

Care has been taken to present everything about tyres; right from their origin to the upcoming tyres in the new digital and sustainable world. In case of anything missing or any deviations found, please feel free to write to me at rajnish@autoscanthetyrehub.com

Thank you.

ACKNOWLEDGEMENT

At the very onset, I must confess, I was thinking of putting up a book on tyres for long, but had this pandemic not happened, this book would not have become a reality anytime soon. Thus, this was a positive side of this pandemic which forced us into lots of lockdowns, people getting enough time for doing things that they always wanted to do.

I cannot thank enough, my teenage daughter, Palak, who was my motivating star and my edit master of this work.

Any long-term work does require family support and I must thank my wife, Seema; my parents, Jagdish and Pushpa Agarwal, my brothers Asish, Satish and my brother-in-law, Sarat, for their inputs.

I'm also thankful to our office managers, Mr Gouri Shankar Patnaik, Mr Chandra Sekhar and Mr Nissank for their valuable suggestions from time to time.

I will also remain indebted to Mr Tarani from Michelin and Mr Mayank from JKTyre for their guidance and support.

I would also like to thank my publishers 24by7 Publishing for being the first publishers to reach out to me, for giving a definite shape to my work and for making this work reach all of you.

Finally, I would like to thank all those who directly or indirectly have helped me in completing this work.

FOREWORD

I, on behalf of JK Tyre & Industries Ltd., congratulate Rajnish Agarwal, who has been associated with us for the last 21 years as JK Tyre Steel Wheels partner, for bringing out a book on tyres to spread awareness about the product and their care. I wish him all success for 'Tyres over a cup of coffee' titled book.

Mr Mayank Pandey, Regional Manager, JK Tyre & Industries Ltd. Cuttack Region

Mr Rajnish Agarwal is passionate about tyres and our Michelin Brand. He is associated with us as a Michelin Tyres & Services Partner for the past 2 years.

The book, "Tyres over a cup of coffee" is an effort towards spreading awareness on tyres and why it needs to be cared for regularly. A tyre is the only link of the vehicle with the road. It is a must-read book.

Mr Tarani Patra, Passenger car & Two-wheel, Michelin India (P) Ltd.

Rajnish Agarwal has been associated with us for almost 15 years and is also a Bridgestone Select dealer now. His vast knowledge of tyres and their services has helped him to write and publish this special book on tyres. I wish him success in all his endeavours.

Mr Viky Singh Sarhadi, Bridgestone Tyres Private Ltd.

PREFACE

Tyre says, “Know me more”! “Give me my due importance”!! “Your life is in my hands”!!!

Written with a purpose to make the tyre users and auto enthusiasts aware of all information on tyres and to show why tyres are so important for our safety and an integral part of our vehicle.

Right from the origin of wheels; the origin of the name itself; how it evolved over generations; the biggest tyre companies of our nation to the biggest global tyre companies.

The next phase is very important as it comprises all about tyre care and its maintenance; how to choose new tyres; tyre care equipment that one should keep in the vehicle and much more.

Other important aspects like tyre recycling and tyre retreading are also mentioned in detail.

You also get to know about the upcoming tyres as we move ahead, with airless tyres by leading tyre companies expected very soon, which will change the way we travel. They will make our journeys trouble-free and safe. This will lead us all into a better and sustainable world.

So, wish you all a heavy dose of ‘the tyre.’

CONTENTS

CHAPTER 1

TYRES, OVER A CUP OF COFFEE

'Giving up on your goal because of one setback is like slashing your other three tyres because of a flat'

~Christine Kane

Imagine yourself alone one night on a self-drive mission crossing a beautiful and inviting jungle, raining heavily, with soft music playing inside, as if from elsewhere and you thinking to oneself of it being such an amazing drive when you only wish you were not all alone in your car and then suddenly comes a 'dhud' sound...your car stops with a big jerk!!

You get down, and to add to the horror...you see a flat tyre which you had not encountered any time before and then you come to think, if only you cared for your tyres, more. You wish you had checked your tyres before the self-drive mission, as there was no tread left on that particular tyre and to make the scene scarier, there was no spare tyre in your car to change to.

The beautiful drive changes to a pathetic nightmare.

Had the ride been with your loved ones and had it been a tyre burst, the result could have been fatal. This is how important, a tyre is, for your vehicle and your safety, which is why we need to pay attention to them on a regular basis.

Tyres & Coffee: What's the connection?

At the very onset, most of you would be wondering if there's any link between the two topics that the title of the book conveys, that is, Tyres and Coffee.

Well, the concept in my mind was that whenever you visit a tyre showroom for tyre replacement or car showroom for servicing, you generally spend an hour or couple, then why not have a book to grasp some information on tyres while you happily sip over a cup of coffee which most of the showroom and service centres offer out of courtesy to enhance customer satisfaction.

After all, we have just seen how important a tyre is and some knowledge on this would help us not only in choosing the right tyres for our vehicle but also in taking better care of them which ultimately will keep us safe and prevent any risk of an accident arising out of our negligence. This will go a long way in reducing the number of deaths that occur due to road accidents as a result of tyre bursts, almost every day of the year.

A few car showrooms and tyre showrooms too, tie-up with coffee brands to offer best of both and to make customers spend time with them so that the sale deal is finalised. Most of the car showrooms today do have a special coffee or refreshment zone. Some even do create a children play area for customer engagement. A few even offer free driving classes once the car is sold.

Did you know? Toyota, in Japan, has a dual concept of I Drive to go with coffee to lure in the young crowd for car rentals who are not yet ready to go for a purchase eyeing future sale.

A tyre showroom in Brunei offers a $10 voucher for every tyre purchased. All these are selling concepts towards enhancing footfalls resulting in a rise in sales.

Tyre says,

"Know me more"!

"Give me my due importance"!!

"Your life is in my hands"!!!

Our movement over long distances; be it on roads, railways or even via airways for that matter are all dependant on tyres and wheels. And we rarely do care while changing into a fresh set of tyres for our vehicle. It's just a round black rubber item that all looks the same to innocent users. But there is a whole lot of science that is involved in making those round rubber items and thus we need to understand that to some extent to choose the tyres that would best suit our vehicle.

By going through this and subsequent chapters, one would be able to know in detail;

- How to know, when to replace your vehicle's tyres?
- How to choose the best tyres for your vehicle?
- Which wheel to go for; Alloys or Steel Wheels?
- How to change a flat tyre?
- Tyre related items that one should have in the car.
- What happens to your car tyres after you discard it?
- Tyre biggest brands and their revenues.
- The upcoming innovative tyre concepts
- The importance of correct inflation and other regular tyre maintenance services to enhance tyre life and safety.

Find all the above and much more about this not so appealing but the most integral part of our vehicle, the tyres!

What is a Tyre?

Tyre, for a common man, is a round-shaped rubber item that is the link between your vehicle and the road that carry the load of the vehicle. Its main raw material is natural rubber, also called latex, synthetic rubber, a host of chemicals and other steel ingredients for strength depending upon the type and purpose for which the tyre is being made.

As per Wikipedia, 'A tire (American English) or tyre (British English) is a ring-shaped component that surrounds a wheel's rim to transfer a vehicle's load from the axle through the wheel to the ground and to provide traction on the surface over which the wheel travel'.

The word tyre has been derived from the French word 'tirer', meaning, 'to pull'. The story of the name Tyre reaches back through Scottish history. The name evolved for a person who worked as a carpenter. The Gaelic form Mac a t-saoir means son of a carpenter or McIntyre. Thus, the name tyre, has its origin on a surname and even today there are a good number of people with the tyre, as a surname, one such name is John Tyre, the most widely used one.

On a funny note,

A tire is a ring-shaped rubber item in a vehicle that tire-lessly tires to reach a destination without [actually given an option for] tir-ing and doesn't easily re-tire.

Importance of Retailing and Merchandising:

For shopping in this digital era, we look not only for convenience but choices as well. We prefer to buy the best of products from exclusive showrooms or shopping malls that offer us variety with innovative merchandising. Be it clothing, shoes, or anything for that matter, even books. Car showrooms, these days are huge in their size and leave no stones unturned when it comes to hospitality to let their customers have an unmatched buying experience. Most of them even have a special coffee zone to make their customers spend more time with them. And as they say, 'Important decisions are taken over a cup of coffee'. So, why should the tyre showrooms be left far behind? After all, tyres are the most integral part of a vehicle which acts as a saviour when cared for or a destroyer too, when ignored.

Tyre Franchise concept store

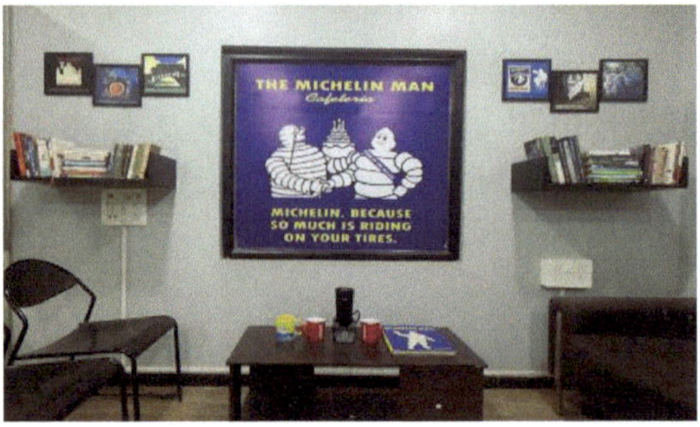

Cafeteria at a tyre showroom

Coffee zone of a car showroom

We have lots of branded tyre showrooms, also called Franchise, which offers the consumers a rich tyre buying experience in a cosy environment. They not only offer you different types of tyres to choose from but also a host of services right from the automated fitment of tyres to 3d wheel alignment, other tyre-related and car-related services too.

This particular Michelin showroom has a unique concept of displaying all other brands names too in their franchise store along with a priority to their own brand so as to offer choice and increase footfalls of tyre buyers to their showrooms.

These highly attractive branded tyre showrooms with service set-ups have made tyre shopping interesting and an exclusive buying experience for vehicle owners.

Tyre display at a brand concept store

The showrooms offer a wide variety of tyres for a particular vehicle along with experts who advise depending upon the driving terrains and climatic conditions.

There are also numerous multi-branded showrooms that offer more than one brand to choose from. Almost, all top tyre brands like Michelin, JK Tyre, Bridgestone and others too have their franchise showrooms in all important locations of the country and worldwide so as to improve their reach and provide tyre solutions to the vehicle owners.

The tyre companies are even moving one step ahead by starting tyre mobile solutions by offering dial-a-tyre service and mobile van service too which has all tyre service machines installed over a van for roadside assistance and home service, just as RSA or roadside assistance, provided by the car industry.

Tyre service on the move

Innovative Tyre merchandising or product displays at showrooms too play a vital role. It is the way the products are showcased at the showroom along with the literature that best describes that particular product, sometimes with special lighting.

Product specific tyre display

The above picture is a Ranger series tyre display of an all-terrain tyre from JK Tyre which is showcased as the tyre has reached the top of a rocky mountain.

Internet connectivity and mobiles are changing the way we live and shop. You as a buyer generally do most of the research at home over the internet and then walk into a showroom generally for the final purchase. Thus, this has led to price competitiveness and hence all the sellers offer the best price with the best possible service. All this to lure you into their showroom for the first time, and once you walk in, then they try their best to make you a regular customer which can only happen if you are thoroughly satisfied with the product, services and hospitality provided. 'Once a customer, always a customer, should be the motto for the sellers and then you won't mind walking again as long you are happy and satisfied.

These days there is also a fashion of upsizing of tyres so as to enhance the overall look of the vehicle along with the feel of the vehicle too.

Wide grip racing tyres

By feel, I mean the grip and control of the vehicle as larger and wider tyres offer more grip and thereby improve the ride and handling quality. Some go for white lettered tyres and tyre stickering too to enhance the overall aesthetics of their vehicle.

Thus, tyre shopping is a very important aspect of your life today. After all, your safety and drive quality depend on the quality of the tyres.

Generally, tyres are not given the importance that they should be given.

An advertisement specifying braking ability

We just tend to buy any tyres that are available nearby but we should carefully choose tyres as per our driving habits and the terrains that we mostly drive on. As much as 20% of all road accidents are because of tyre bursts and hence proper care and maintenance are of utmost importance. Also, find all about tyres, right from their origin to innovations till date with special reference to the world's leading brand. And moreover, since the pandemic is on a high, and has slowed down all economic activity to a certain extent, even production of tyres as the sale of the auto industry has also been restricted. And the demand for tyres is directly dependent on the performance of the auto sector as they are the primary customers for tyres as OEM (Original Equipment Manufacturers) such as Maruti Suzuki, Hyundai, Tata Motors, to name a few.

Waiting for this pandemic to bow down completely, so as to see the boom of economic activity again and thereby the auto industry and the tyre industry.

CHAPTER 2

EVOLUTION & HISTORY OF TYRES

Will you believe if I say humans started to think of moving things as early as 3500BC? Yes, it's true.

Almost all of us would be interested in knowing as to actually how did this tyre evolved into being the tyre that it is today which we use on a very regular basis as it is the only medium of transportation on land for long distances along with railways.

To understand the origin and the subsequent development of tyres that we see today, we first need to understand the origin of the wheel and its subsequent development as it went on to give birth to tyres which along with its rim has gone to reduce rolling resistance to a great extent.

Source: evolution treadwright.com

Did you know? The first wheel was a round rock with a hole in the centre to put up an axle to carry loads but they wear and tear in an uneven way and it was difficult to carry on for long distances. Thus, this inspired people to try out logs of trees to roll over heavy objects from one place to another.

We all have seen bullock carts and horse chariots in villages and movies where the wheel is generally a large piece of circular wood with a hub in the centre which becomes its axle used for rolling.

Thus, humans when they experienced that pulling big things was a big task, they started to try out logs of trees to roll larger things as they found out that the rolling resistance is far lower. This gave birth to the first wheel somewhere in 3500 BC, i.e., some 5500 years ago.

The invention of the wheel and axle allowed a cylindrical log to be placed in the centre and gave birth to chariots to be pulled by animals as we so often see them in our mythological video series on television.

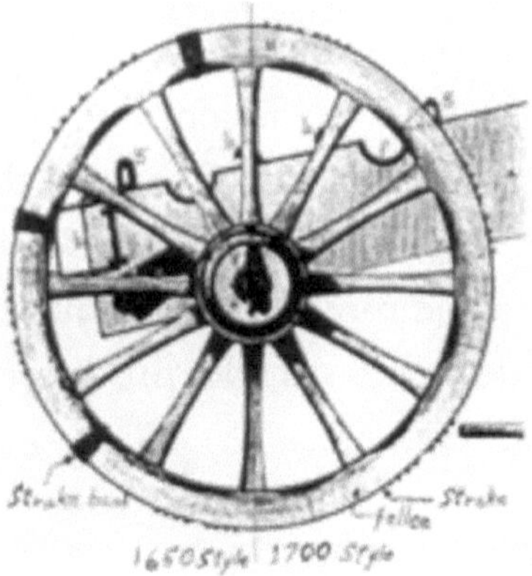

A craftsman, known as Wheelwright was famously recognized as the pioneer of this form of tyres.

Source: cartwheel: tyremarket.com

Then came the spoke wheels which still made the rolling resistance lesser and so forth until finally air-filled pneumatic tyres were developed in the 19th century which we still use today though technically superior and with lower rolling resistance.

Today, tyres can touch speed up to 400kmph which is an indication of its advancements in technology.

Wooden wheels with Leather and Steel

Once the usage of the wooden wheel was widespread, people began to think about reducing the jerks. So, they started to wrap leather around the wheels so as to take up some amount of jerks and make travelling smoother with better traction. Also came iron and steel wheels to lessen down wear and tear. Trains, to date, run on heavy iron wheels but on tracks specially laid down for their movement.

Solid tyres:

Charles Goodyear gave birth to vulcanized rubber and vulcanized rubber helped to invent solid rubber tyres which made movement easier and faster though they were still quite heavier. They were strong and could take cuts and abrasion. Still, smoothness in the movement was missing as it could not take many jerks that the vehicle produced. Then came the ultimate pneumatic tyres.

Wheels through the ages

The wheel has been used extensively and improved upon throughout history, but how have humans harnessed its practicality?

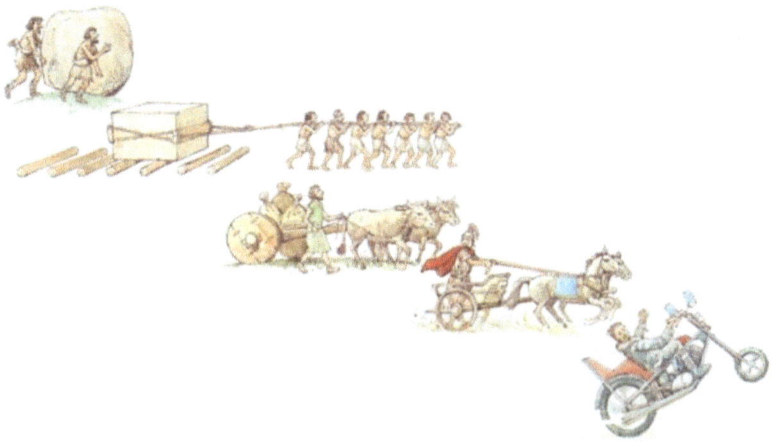

Source: wheels ages: historyanswers.co.uk

Pneumatic Tyres:

In 1847, Robert W Thompson, a Scottish engineer, created and patented the first Pneumatic tyre though it didn't get into production at all.

But in 1888, John B Dunlop created the first successful pneumatic tyres which went into production and were successful and it changed the rolling resistance to a great extent, the technology that we still use today.

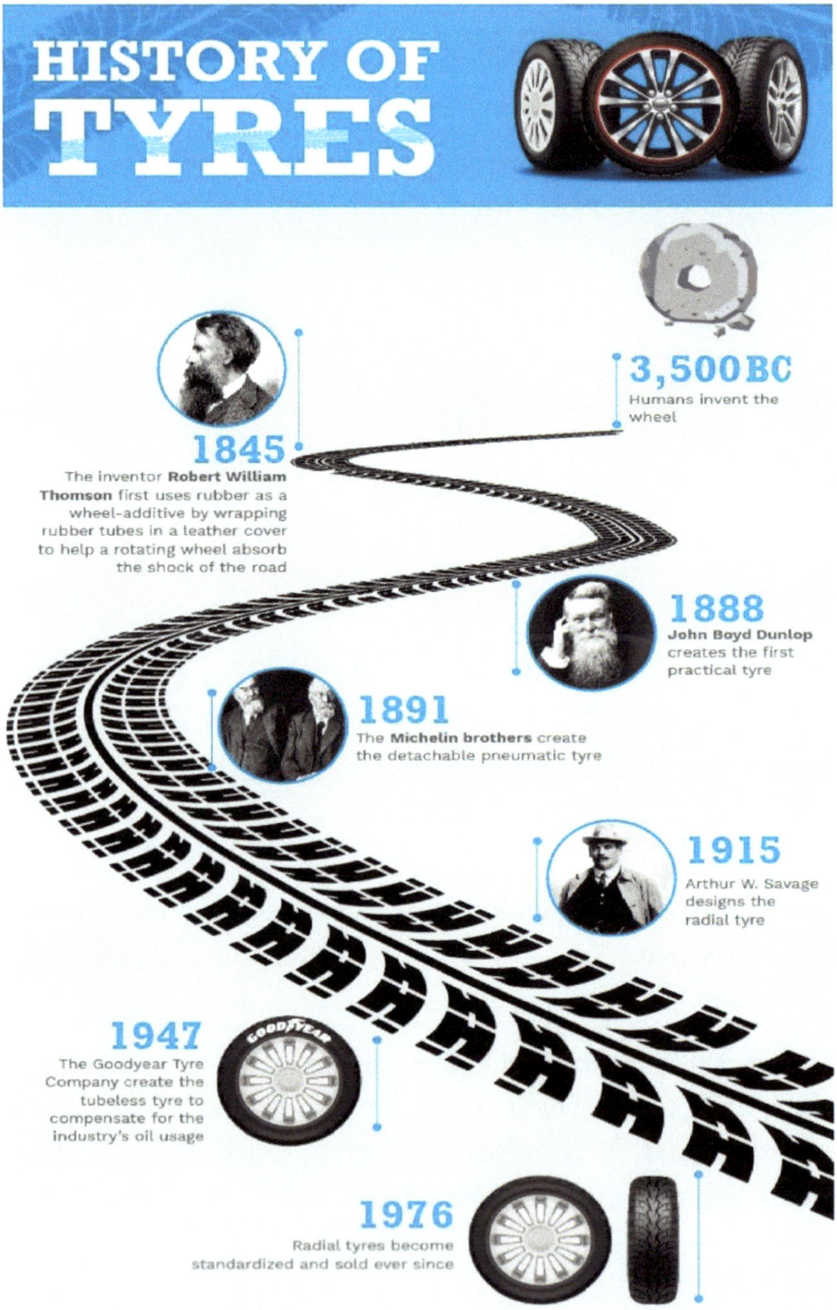

Photo Credit: whattyre.com

Radial tyres:

Radial tyres are the tyres we still use today; almost 100% in cars and 60% in commercial vehicles. Michelin invented radial tyre technology after World War II. This technology though did not catch up until the 1970s but as of now, this is the most widely used tyre especially in cars. Large commercial vehicles in most countries still use cross-ply bias tyres as they carry heavy loads on rough roads better.

Radial tyres use steel belts for better strength and it greatly reduces the chances of punctures.

Radial tyre	**Bias tyre**
Comfortable drive	bumpy ride due to stiffness
More fuel-efficient	less fuel-efficient
Expensive	Cheap
Excellent traction	Weak traction
Increased durability	Less durable
Can be difficult to repair	Easy to repair
Susceptible to damage	Less susceptible to damage

Since the radial tyre was first introduced to the world by the tyre giant Michelin in 1946 and also because the company is as old as the tyre technology in itself along with being the largest tyre manufacturer and an iconic tyre brand, let us have a look at its evolution as well to understand the history of tyres in a better way.

Michelin Tyres Evolution:

Michelin tyres is a French multinational tyre manufacturing company based in Clermont–Ferrand, region of France. It is the largest tyre manufacturer in the world. Apart from the Michelin brand, it also owns a few other brands like Kleber, Goodrich, Camso among other brands. Michelin is also notable for its red and green travel guides, its road maps, the Michelin stars that the red guide awards to restaurants for their cooking and for its company mascot, Bibendum, colloquially known as the Michelin Man. Bibendum, itself is more than a hundred-year-old mascot and one of the most recognised mascots in the world and this has helped Michelin to promote itself as a premium brand along with its quality performance. Michelin has just become a 132-year-old company which was established by Edouard and Andre' Michelin in 1889.

Its history is almost as old as the history of the tyre itself and most of the innovations are led by Michelin with its strong research and development facilities throughout the world.

Thus, it would be unfair to not have a special focus on this special brand which has made a tyre mascot named Bibendum, as famous as the tyre itself.

- Born in 1889 in Clermont-Ferrand, France
- Established by Edouard and Andre Michelin
- In 1895, develops the first car
- 1898, Michelin man, Bibendum was born
- 1900, the first Michelin guide was published
- In 1910, road maps were published
- 1930, a tyre with a built-in-tube introduced
- 1946, radial technology introduced
- 1952, radial technology for truck tyres

- 1981, radial tyres for aeroplanes
- 2000, Bibendum becomes the most recognised mascot
- 2004, introduced tweel technology tyres
- 2007, mems tyre monitoring system launched
- 2019, UPTIS, airless tyre introduced for cars
- 2019 sales revenue was $23.65 US Billion
- CEO: Florent Menegaux since 17th May 2019
- Present in 170 counties and has 123400 employees
- 71 production facilities; produces 170 million tyres
- Produces tyres for all segments including NASA too.

Important Dates in Michelin's Evolution

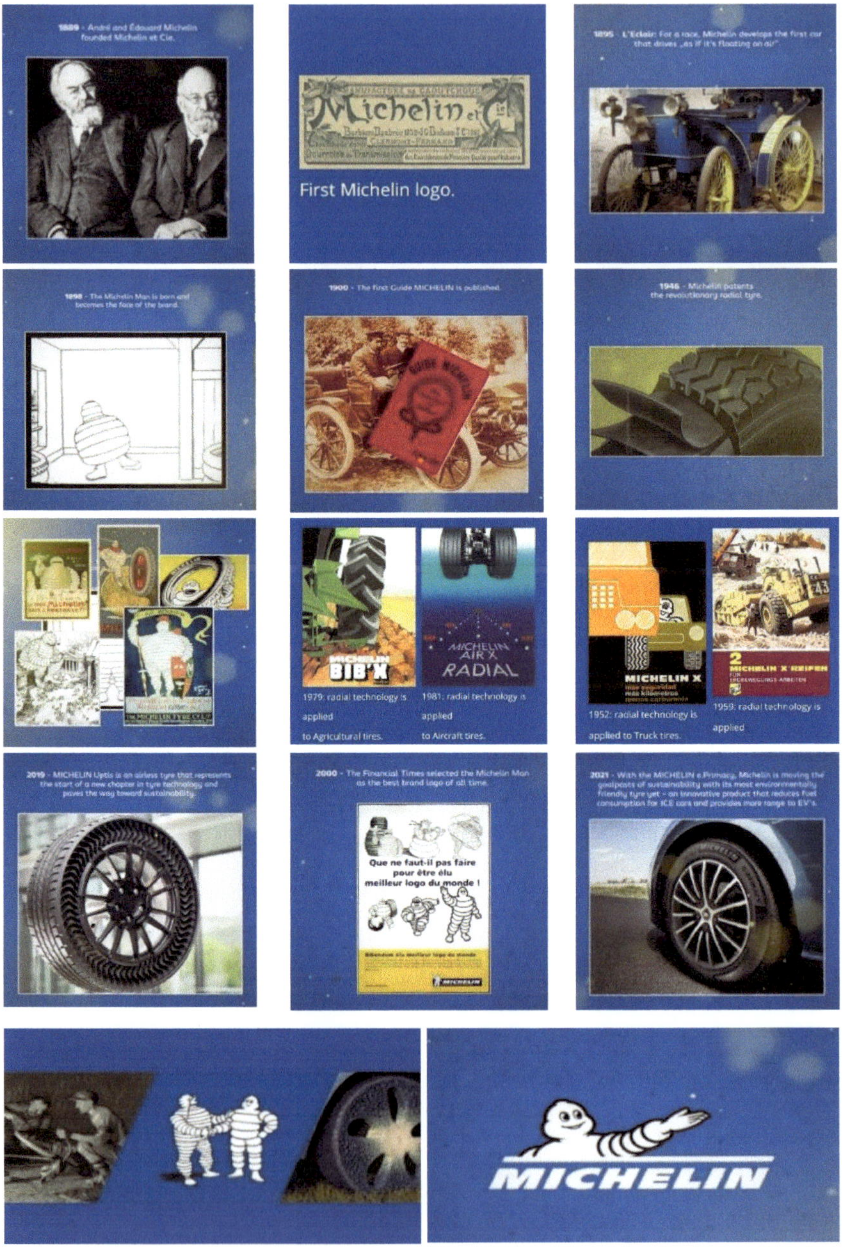

Photo Credit: Michelin.com

CHAPTER 3

STRUCTURE OF A TYRE

The radial tyre consists of 9 main parts:

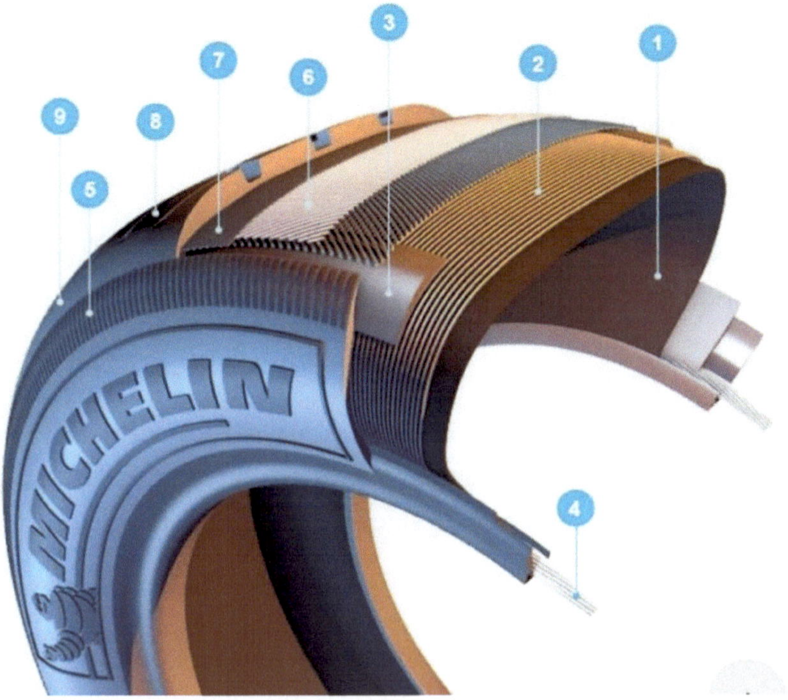

Source: structure: Michelin.com

1. Inner liner:

   An airtight layer of synthetic rubber (the modern equivalent of an inner tube).

2. Carcass ply:

   The layer above the inner liner, consisting of thin textile fibre cords bonded into the rubber. These cords largely determine the strength of the tyre and help it resist pressure. Standard tyres contain about 1400 cords, each of which can resist a force of 33lb.

3. Lower bead area:

   This is where the rubber tyre grips the metal rim. The power from the engine and braking effort is transmitted from the rim of the tyre to the contact area with the road's surfaces.

4. Beads:

   They clamp firmly against the tyre's rim to ensure an airtight fit and keep the tyre properly sealed at the rim.

5. Sidewall:

   It protects the side of the tyre from impact with curbs and the road. Important details about the tyre can be found on the sidewall, such as tyre size in details and speed rating.

6. Casing ply:

   It largely determines the strength of the tyre. It's made up of very fine resistant steel cords bonded into the rubber. This means the tyre can resist the strains of turning and doesn't expand due to the rotation of the tyre. It's also flexible enough to absorb deformations caused by bumps, potholes and other obstacles on the road.

7. Cap ply:

   This is also called the zero-degree belt. This important safety layer reduces friction heating and helps maintain the shape of the tyre when driving fast. To prevent centrifugal stretching of the tyre, reinforced nylon-based cords are embedded in a layer of rubber placed around the circumference of the tyre.

8. Crown piles or belts:

   They provide a rigid base for the tread.

9. Tread:

   It provides traction and turning grip for the tyre and is designed to resist wear, abrasion and heat.

The importance and the science of tread design.

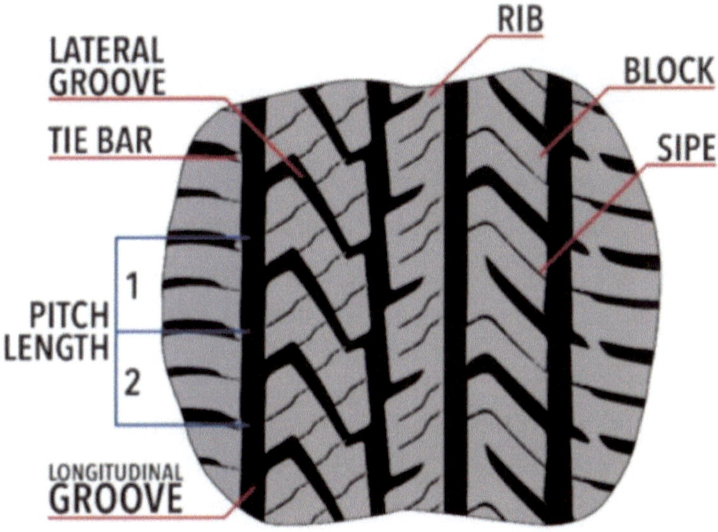

A common highway tyre tread pattern

Have you ever wondered about the science behind a tread design or pattern? Why some tire tread designs are sleek, some symmetrical, some directional and some a mixture of all? Why do all those small cuts and gaps exist in the way that they do?

Well, the only contact part of the tyre with the road is the tread and thus its design and pattern play a major role in determining how well a particular tyre performs.

It is the tread that bears all the abrasion of the roads at various speed levels, takes care of all the small stones, water bodies, sand and what not. Hence the tread and its grooves should be prepared for all these to assure us safety at the time of actual performance.

Engineers spend hundreds of hours designing the tyre pattern for optimum performance depending upon the main functions that it needs to perform.

Though each of the above nine parts as seen in brief goes into making a tyre and have their own importance but it is the tread that comes into actual contact with the road surface and that is why its design becomes that much more important and research worthy.

As there are thousands of varieties of tyres, there are again thousands of tread designs each having its own unique features to provide certain functions that the tyre engineers intend them to perform.

As layman buyers, we hardly ever see the tread design before buying a tyre and a seller too, other than in authorised branded stores, seldom tries to explain the unique tread design and its functions.

But if you happen to ask a tyre engineer, a lot of scientific formulas goes into designing a tyre tread as it is required to provide the basic function of traction on the road surface and its design is very much dependant on the surface that it needs to run on, most part of its life.

For example, a high-performance summer car tyre has plain rib grooving for tearing the air at high speed.

A tractor rear tyre has big angular and wider tread for traction on muddy fields; similarly, an off-roader has a stony protruding sort of tread design for better traction on sand and stones.

Different tread patterns based on season and surface:

Hence, a lot of scientific research goes in for designing various treads and each tread design is meant to perform a certain set of functions while at work on the road or elsewhere.

Functions of tread:

1. Traction: This is the basic but most important function that a tread pattern performs.

   A better or superior grip is an important tyre or tread function that many tyre manufacturers intend to convey to their customers. A lot goes into designing a tread so that it provides a superior grip even at speeds of 200 to 300 kilometres per hour. Hence tyre is required to mention the speed limit that it can withstand on the sidewalls. The temperature and the noise level should not rise beyond a certain level even at that speed to ensure safety.

2. Braking: This function is again one of the most important functions that the tread performs. Shorter braking distance is what the tyre manufacturers design their tyres for so that in case of emergency, the tyres can save lives.

   The tread design which has certain grooving patterns traps the air to stop at a shorter distance.

3. Cornering: Not only braking but cornering and turning functions too are very much dependant on the tyre as a whole but again is the tread that helps in a faster response.

4. Superior wet grip: Wet grip on the road becomes more important as tyres with lower quality tread slip and it can prove fatal. Thus, the tread design helps to tear down water bodies on road thereby providing better traction and it should avoid hydroplaning where the tyre tread loses its grip on shallow water bodies on road especially at a higher speed and it basically floats which may result in fatal accidents.

5. Noise Reduction: This is one more aspect that depends on the tread design again. Tread patterns are generally designed to keep the noise to the lowest so that the car users are not disturbed thereby offering a safer and pleasant drive.

   The level of noise and the wet grip is going to be monitored and the government is making it mandatory for tyre manufacturers to mention these on the tyre sidewalls.

   Thus, we need to choose our tyre carefully as per our usage, the surface that we would generally drive on, and the climatic conditions of the place where the vehicle would mostly run.

CHAPTER 4

TYPES OF TYRES

Tyres come in a lot of sizes depending upon the vehicles and their application, depending upon climatic conditions such as for icy roads and extremely hot climates, depending upon the terrain such as off-roader, all-terrain.

Thus, the types of tyres depend upon numerous factors and so thus its construction for optimum results.

The different types of tyres are best illustrated by the table

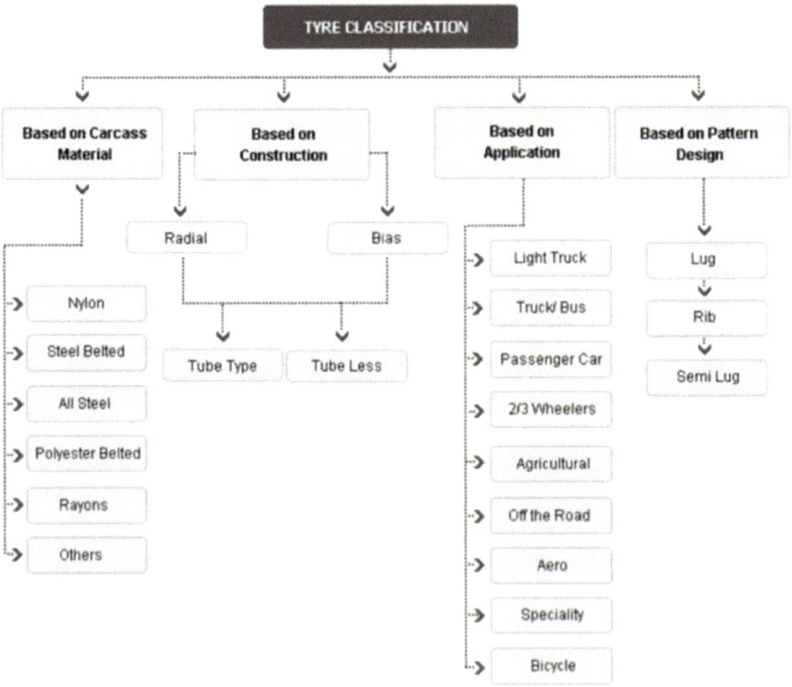

We shall look in detail, the widely used ones today.

Source: carwale.com

According to the material used and construction:

1. Radial Tyres:

Radial Tyres today dominates the Indian market which was not the case in the year 2000 when radialisation in India had just started and only a few cars used to have radial tyres fitted in them as they rolled out along with the vehicles such as Tata Estate, Maruti Cars premium models to name a few. Today in 2021, only a handful of Trucks are left which doesn't come fitted with radial tyres from the OEM [Original Equipment Manufacturers] side. Cars are 100% radialised today.

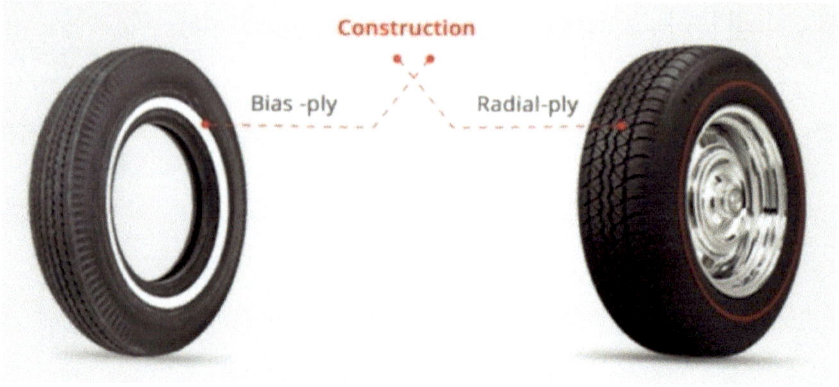

Source: carwale.com

This growth in radialisation in India and most of the countries are because of the following reasons:

a. Comfort:

Radial tyres feature steel belts running to a 90*angle to the tread centre line. These tyres are constructed in 2 parts allowing the tread and sidewalls to function independently. Hence the Sidewalls take lots of jerks allowing the cars to run smoothly without making the passengers feel the jerk. Hence these types of tyres are known for comfort and smooth drive.

b. Puncture Resistance:

Since a number of steel belts are used just below the tread, the chances of punctures are negligible in radial tyres in comparison to Nylon tyres where carcass layers are made of nylon threads.

c. Greater Safety:

Since the sidewalls are flexible, the contact area of tread is larger in radial tyres which gives more grip on the road surface thereby increasing safety along with comfort too.

d. Lower fuel consumption:

The transfer of energy from machine to the road is better and hence there is lower fuel consumption as ultimately the friction between roads and tyres are lesser in radial tyres compared to nylon tyres.

e. Lower heat generation:

In radial tyres, the heat generation at higher speeds is quite less owing to its construction pattern especially because of the sidewalls taking lots of weight of the vehicle though the exposure to sidewalls cuts is high as they are softer compared to a nylon tyre construction.

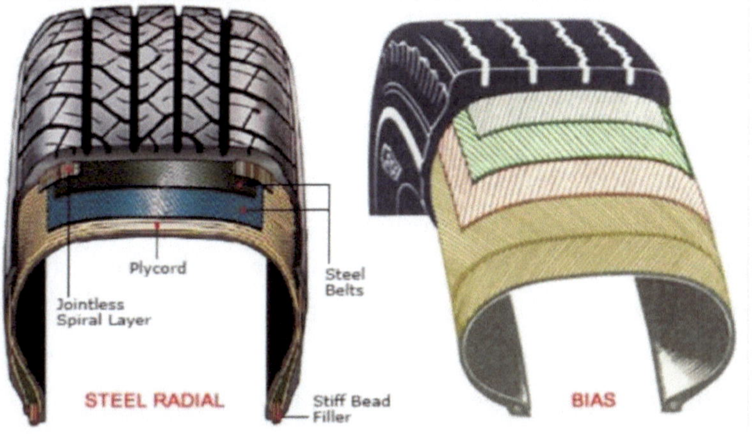

Radial Tyres Vs. Bias Tyres

Source: tyremarket.com

2. Nylon tyres or Bias tyre:

Cross-ply or Nylon tyres consists of carcass layers made from nylon cords. They are placed diagonally against each other. Ply rating in tyres used to come earlier and even does come now which indicates the number of times the nylon cords are laid upon the carcass. Thus, a 10-ply tyre will have more strength than the 8-ply rated tyre.

The nylon tyre belts run at 45-degree angles against the tread line. The multiple overlapping piles in these tyres connect the sidewall and the tread.

Thus, this results in stiffer sidewalls as compared to radial tyres. Though it offers better rides on rough roads and is less costly too compared to radial tyres, these are generally used in commercial segments as these are cheaper and generally cost factor is a major factor in tyre purchase.

But in reality, even if these tyres are approximately 20 to 25% costlier than the nylon counterparts, they are better in the long run as the life of the tyres are longer by approximately 50% along with better per kilometre fuel mileage. There is lesser wear and tear because of a lesser amount of jerk to the vehicle and this results in lesser fatigue for the drivers too.

3. Tubeless tyres:

This is relatively a new concept especially in radial tyres where there are no separate inner tubes but an inner liner that prevents leakage of air in case of a puncture and as the leakage of air is very slow, the vehicle user can reach a repair centre even if they are far away. The air is in between the tyre and the rim as compared to a tube-type tyre where the tube is air-filled.

The repair system of this kind of tyres are actually easier and these tyres are being used by the OEMs today and hence in the replacement market too.

Tubeless tyres

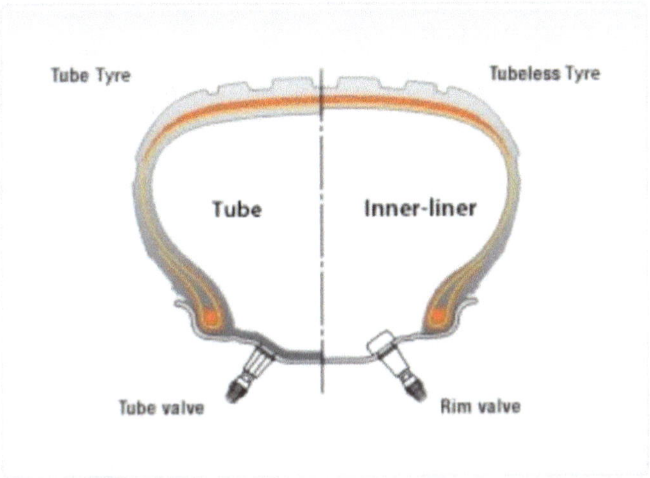

Structural comparison of tube/tubeless tyre

Source: TyreDekho.com

4. Solid tyres:

A solid tyre is an airless tyre. It is not filled with air but is manufactured using layers of rubber that are constructed around a metal frame or wheel structure that can be mounted to a specific vehicle. They can also be made to fit on rims manufactured to support pneumatic tyres.

Although solid tyres predate pneumatic tyres, they still play a major role in the agriculture, industrial and logistics sectors. Solid tyres are extremely stable, economical, puncture-proof and have a high load-bearing capacity in extreme service conditions.

They are best suitable for forklift trucks, airport vehicles and industrial vehicles.

5. Run-flat Tyres:

RFT or Run Flat Tyres are tyres on which you can continue driving even after a puncture and thus you can take time before you reach a repair shop.

In most self-supporting run-flat tyres, the reinforced sidewalls carry the vehicle for a certain distance. The supporting ring-type employs a hard rubber that can support the vehicle's weight in case of air loss. DriveGuard tyres from Bridgestone is an example of a run-flat tyre.

According to seasons:

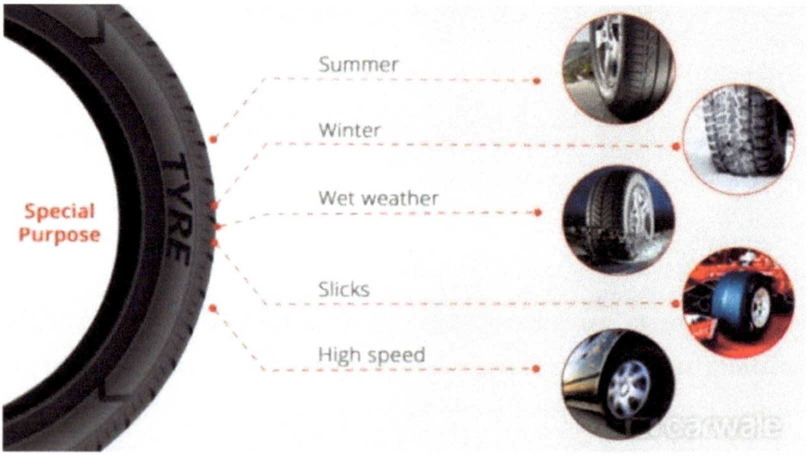

Source: carwale.com

1. Winter Tyres:

Tyres meant for colder temperatures and icy roads are called winter tyres as they provide better traction and where thin metals are fixed for even better traction and to avoid skidding on icy roads or studless winter tyres where metals are not fixed but have special and deeper treads for better grip on winter roads. These are more popular today.

2. Summer Tyres:

High-performance tyres with smaller tread groves for temperatures above 7°C. It has got firmer tread for better grip on dry roads.

3. All-Season tyres:

These tyres perform optimally throughout the year and are the most widely used tyres. They have even tread grooves to cater to all seasons where there are no extreme temperatures.

According to Terrain:

Source: Carwale.com

1. Off-Roader tyres:

These tyres are also called 4*4 tyres as they are better suited to driving off the road, that is in muddy and slushy roads, in sand dunes, in hilly roads and on rocky water beds. These tyres also perform well on proper highways. They have wider tread and deeper grooves. JK Tyre all-terrain Ranger series is an example of an off-roader.

2. Highway tyres:

These tyres are high-performance smooth tyres meant for longer journeys on highways and are among the maximum selling tyres in India and elsewhere except for very cold countries.

3. Snow tyres:

Snowy or winter tyres, as seen earlier too, have deeper grooves for better traction on snowy surfaces and icy roads. They could be studded with metals for extreme conditions.

According to the application:

1. TBR/TBB or Truck Bus Radial and Truck Bus Bias:

Truck and Bus radial tyre is dominated by JK Tyre in India as the company was the pioneer when it came to radialisation be it in passenger cars or trucks and buses.

Apollo dominates the bias market followed by MRF. New trucks have reached up to 75% radialisation as demanded by OE and the replacement market has reached up to 55% in India owing to road conditions in rural areas though there is speedy development of infrastructure.

2. LCV or Light commercial vehicle:

Radialisation is on the lower side as these are again for commercial purposes. This segment is led by MRF followed by Apollo and JKTyre.

3. PCR or Passenger Car Radial:

Bridgestone is the undisputed leader in this segment as most OE have accepted the brand and so it is in demand at the replacement market too owing to its long-life span and comfort. JK Tyre, was the pioneer in this segment and is leading the way presently with smart tyres.

4. Tractor Tyres:

These tyres account for nearly 6-7% market share in India and are dominated by Goodyear though Indian companies Apollo, MRF and BKT follow closely behind. These tyres have very wide criss-cross grooves for better traction on agricultural lands and mud.

5. Two/Three-Wheeler tyres:

India is the largest manufacturer of two-wheelers and is also the largest seller of two-wheeler tyres. 50% is the market share of two-wheeler tyres in volume and 12% is the market share in value. Ceat, MRF and TVS are dominating the Indian market.

6. Aero:

This is a lesser-known segment though has been seen by most.

Aeroplane tyres are only made by MRF among Indian companies which are dominated by Michelin worldwide.

- Aircraft tyres are designed to withstand extremely heavy loads over short distances
- Tread patterns are designed to facilitate stability in high crosswind conditions, to channel water away to prevent aquaplaning and for braking effect.

Largest tyres are 14 feet in height for 250 tonner dumper trucks.

According to pattern/design:

1. Lug:

These tyres have parallel lug tread grooves to withstand heavy loads and are fitted in the rear of the trucks and buses and other commercial vehicles. They account for nearly 60% of the three available patterns.

2. Rib:

These are meant for the front wheels as they are meant to show direction and move forward. They come in various rib patterns like 4 ribs or 5 ribs pattern.

3. Semi-Lug:

These have characteristics of both the above pattern and are suitable for the all-wheel application. These patterns are the least selling tyres of the three.

Type of pattern		Samples	Features	Recommended application
RIB			⭘ Good controllability ⭘ Less side skid ⭘ Lower noise ▼ Poor traction on wet or mud	On road (High-way, City usage)
RIB LUG			⭘ Useable both On and Off road service ⭘ Performance between Rib and Lug	On & Off (Local paved road, Dirt, Gravel road)
LUG	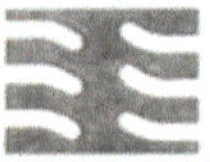		⭘ Good traction on unpaved surface ▼ Noise/Heel and Toe wear	Off road (Dirt, Gravel, Mud)

- Built for performance rather than longevity. Lasts for just 60-100kms
- Tyres are changed before the race gets over at the team's pit stop.
- F1 tyres have nylon and polyester base structures to withstand high speeds and temperature
- Built according to racing tracks- dry, wet and so on.
- Only filled with nitrogen air to withstand huge forces and speeds
- The F1 tyre size is 305/675-13 for the front and 405/675-13 for the rear.
- A set of 4 tyres costs $ 4000 and a team spends up to $20000 on tyres per race.
- Pirelli is the present tyre supplier

8. Bicycle Tyres:

These tyres are the ones that were the first tyres that were introduced to the world. India is a huge bicycle tyre market dominated by Ralco, Syndicate, Bedrock, Metro, GRL (Govind Rubber Limited) among a few others.

9. OTR or Off-The-Road Tyres:

Mining and Construction are the toughest sectors on earth. So are the machinery requirements and tires, off course. They use deep treads for unpaved surfaces such as mud, sand, rocks,etc. Different sizes and tread pattern are used.

Largest tyres are 14 feet in height for 250 tonner dumper trucks.

According to pattern/design:

1. Lug:

These tyres have parallel lug tread grooves to withstand heavy loads and are fitted in the rear of the trucks and buses and other commercial vehicles. They account for nearly 60% of the three available patterns.

2. Rib:

These are meant for the front wheels as they are meant to show direction and move forward. They come in various rib patterns like 4 ribs or 5 ribs pattern.

3. Semi-Lug:

These have characteristics of both the above pattern and are suitable for the all-wheel application. These patterns are the least selling tyres of the three.

Type of pattern		Samples	Features	Recommended application
RIB			◎ Good controllability ◎ Less side skid ◎ Lower noise ▼ Poor traction on wet or mud	On road (High-way, City usage)
RIB LUG			◎ Useable both On and Off road service ◎ Performance between Rib and Lug	On & Off (Local paved road, Dirt, Gravel road)
LUG	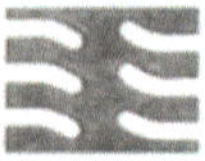		◎ Good traction on unpaved surface ▼ Noise/Heel and Toe wear	Off road (Dirt, Gravel, Mud)

TYPES OF WHEELS:

Steel Wheels or Alloy Wheels: Which is better?

Steel wheels as the name suggests are made of steel and are the most widely used type of wheels in the auto industry. The alloy wheel on the other hand is a mixture of metals like aluminium, nickel, magnesium and a host of others to keep it light and strong in comparison to the earlier ones.

These days, most of the luxury cars and the higher versions of the top-selling cars too comes fitted with alloy wheels for their aesthetic value and also because they are faster.

Steel Wheels:

PROS:

- Cheaper: costs 70% lesser than alloy wheels
- Durable, rigid and stronger and thus comes with most SUV's
- Repairing a steel wheel is cheaper and easily possible

CONS:

- Weight: it's a bit more and hence heavier on suspension and less fuel mileage
- Looks: It doesn't look sportier with a wheel cap
- Rusts: It does rust and that makes its strength less over time

Alloy Wheels:

PROS:

- Aesthetics value: Alloy wheels definitely looks sportier and enhance overall visual appeal.
- Lighter: They are lighter and hence will improve fuel economy and overall vehicle maintenance cost.
- Better Handling: Since they are lighter, the vehicle handling is better.

CONS:

- Expensive: when it comes to cost, they are very expensive
- Durability: in case of a heavy impact, they bend easily and are not easily repairable.

As per the above pros and cons that we have seen, there is not a huge difference in terms of performance and hence even OEMs chose to go for both variety for every model of car that they make to offer choice to their customers. Thus, it's more of a visual appeal and alloy wheels are a must for luxury cars.

CHAPTER 5

TYRE MANUFACTURING PROCESS

A tyre is a complex combination of components made from numerous ingredients. As per Michelin, tires are a treasure trove of technology. But how do we get from the rubber tree to the road? Tyre production is a multistep process.

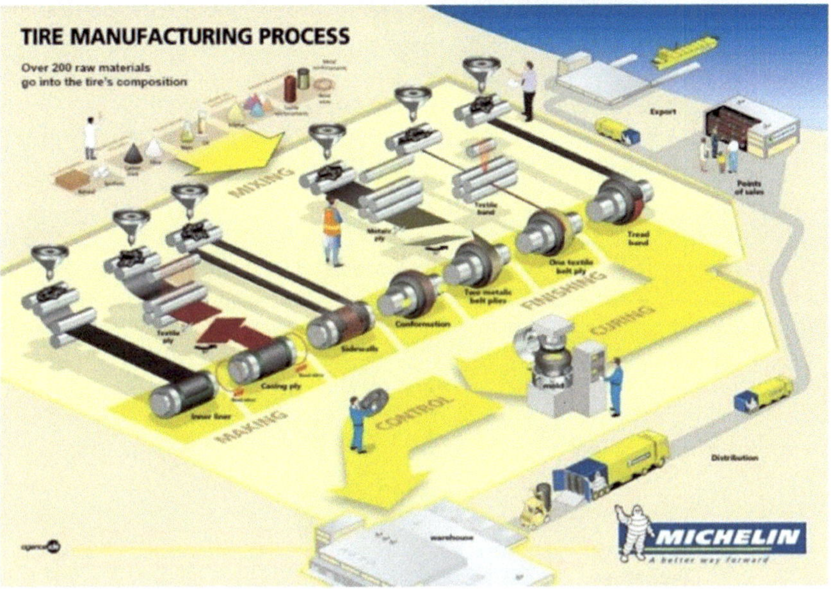

Source: michelin.com

1. Understanding through research:

   Tyre companies study peoples' tyre usage and driving habits to make sure that their tyres meet everyone's needs. They take pride in their quality and hence there is no compromise on research and production process.

2. Developing and mixing materials:

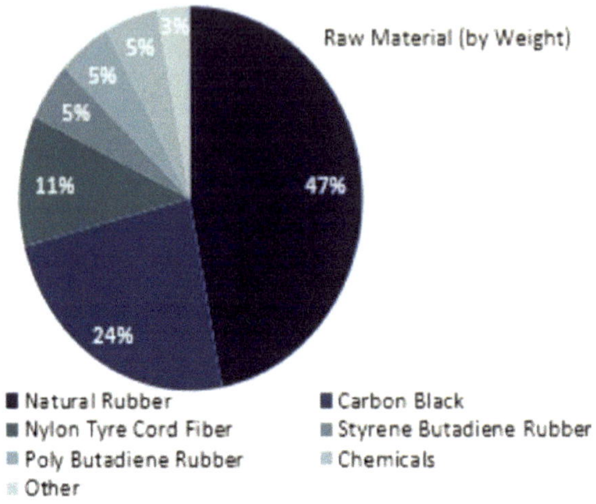

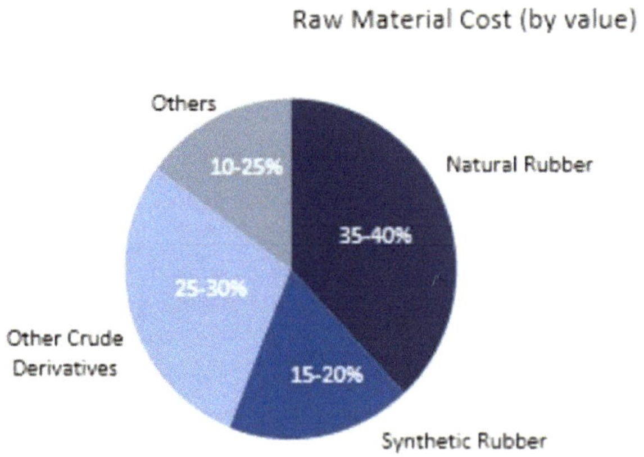

Data Credit: Alphainvesco.com

Over 200 ingredients go into making a tyre. They play vital roles in safety, fuel efficiency, performance and eco-friendliness.

These components fall into 5 groups

- Natural rubber: The main component of the tread layers
- Synthetic rubber: Is used in polymers found in crude oil. Part of the treads of car, van and 4*4 tyres
- Carbon black and silica: Used as a re-enforcing agent to improve durability.

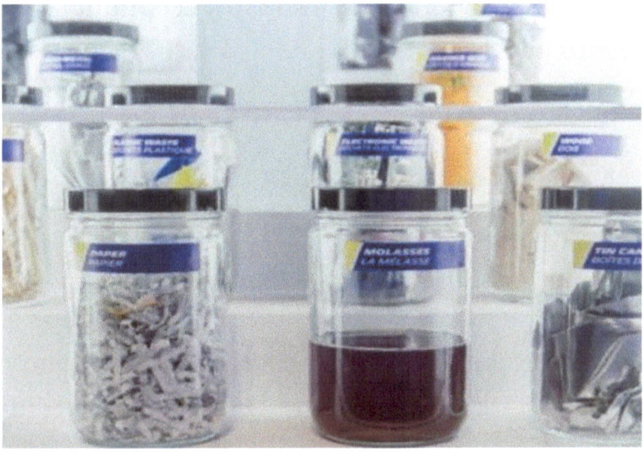

Source: Michelin.com

- Metallic and textile re-enforcement cables: The 'skeleton' of the tyre, forming the geometric shape and providing rigidity.
- Numerous chemical agents: For unique properties like rolling resistance or ultra-high grip.

3. Designing:

   Tyre companies create many different designs and use simulations to test and select the best tyre concepts to be developed.

4. Manufacturing:

   Tyre companies expertly build each tyre through both handmade and automated processes. When necessary, they invent their own machines to reach their goals.

   a. Steel chords:

      Pre-treated steel chord supplied on wire spools is fed into a calendar where it is embedded in one or more layers of rubber. This results in a continuous sheet of chord and rubber.

   b. Tread:

      The rubber material that has been blended in a mixer is now ready to be made into the tread.

   c. Textile chord:

      A multitude of textile threads is fed into the calendar by a large roller device and embedded in a thin layer of rubber.

d. Steel bead:

The core of the bead is made up of many ring-shaped steel wires. Each of them has its own rubber coating. This loop is then fitted with a rubber apex.

e. Sidewall:

Sidewall sections cut to suit the particular tyre size are turned out with the extruder.

f. Inner liner:

A calendar forms the airtight inner liner into a wide thin layer.

Now, the tyre building can begin. Various semi-finished products come together on the tyre building machine and are assembled into what is known as a 'Green tyre'.

5. Vulcanization

It's also called the curing press. Here, the tyre receives its final shape after being vulcanized for a certain length of time at a certain pressure and temperature. During this process, raw rubber becomes flexible, elastic rubber. Curing press moulds are engraved with the tread pattern and sidewall markings.

6. Quality control:

Quality control is not only an end step. It is measured throughout the process. Generally, 1 billion miles worth of tyre testing is done per year – the equivalent of driving around the planet 40000 times.

CHAPTER 6

READING A TYRE SIZE AND UPSIZING

Tyre Sizing and Upsizing or "The Mathematics of Tyres"

As shown in the picture below, the size of the tyre is written on the sidewall of the tyre which denotes its size in detail along with the speed rating and manufacturing code with month and year.

Let's take the size shown 195/55 R 16 87 V which is meant for Hyundai i20 as supplied by the Hyundai company.

In this case, 195 stands for the sectional width of the tyre in millimetres.

Thus, the wider the tyre, the better the grip with the road.

55 is the aspect ratio of the tyre as a percentage of sectional width that is 55% of 195mm, that is, 107.25mm is the height of the tyre and hence the total height of the tyre will be 107.25 +107.25mm + 16 inch or 406.40mm or total height of the tyre is 620.90mm as against 195mm of nominal width. Hence a bit of mathematics is involved over here.

R stands for radial construction of tyre and 16 stands for the diameter of the inner rim in inches.

87 stands for the load capacity of the tyre in kilograms.

Each number is given a special load index that the tyre can withstand.

V stands for speed rating that the tyre has been tested to withstand at full load. In this case, it is 118milesperhour.

Along with the manufacturing week and year, the country of make is also mentioned in the sidewall.

The maximum load that the tyre can carry is also mentioned along with the maximum air pressure that it can withstand.

And how can we forget the brand name of the tyre which is very prominently mentioned along with the model's name of that particular tyre? For example; JKTyre and Ranger series.

Tyre Load Index Chart

Use the table below to find the load and speed capacity from the markings on your tyres:

Load Index	Load (kg)	Load Index	Load (kg)	Load Index	Load (kg)
62	265	88	560	114	1180
63	272	89	580	115	1215
64	280	90	600	116	1250
65	290	91	615	117	1285
66	300	92	630	118	1320
67	307	93	650	119	1360
68	315	94	670	120	1400
69	325	95	690	121	1450
70	335	96	710	122	1500
71	345	97	730	123	1550
72	355	98	750	124	1600
73	365	99	775	125	1650
74	375	100	800	126	1700
75	387	101	825		
76	400	102	850		
77	412	103	875		
78	425	104	900		
79	437	105	925		
80	450	106	950		
81	462	107	975		
82	475	108	1000		
83	487	109	1030		
84	500	110	1060		
85	515	111	1090		
86	530	112	1120		
87	545	113	1150		

Speed Index	KM/H	MPH
F	80	50
G	90	56
J	100	62
K	110	68
L	120	75
M	130	81
N	140	87
P	150	93
Q	160	100
R	170	106
S	180	112
T	190	118
U	200	124
H	210	130
V	240	149
W	270	168
Y	300	186
(Y)	300+	186+

Load Index

Speed Index

Data Credit: Michelin.com

Upsizing or Plus Sizing methods:

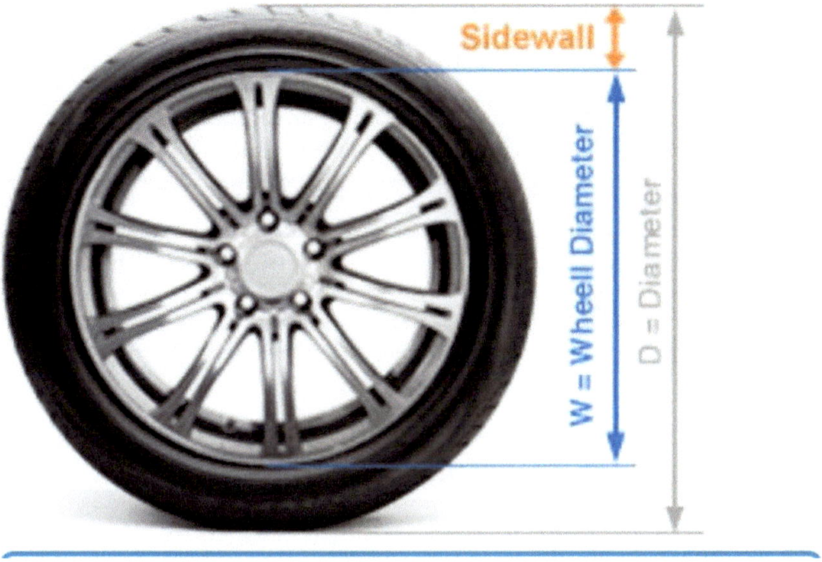

- Plus Zero: Utilizes the same wheel diameter as the OE tyre but combines it with a tyre that has a larger section width and a smaller aspect ratio than the OE tyre.
- Plus One: Utilizes a one-inch larger diameter wheel in combination with a tyre that has one step lower aspect ratio than the OE tyre.
- Plus Two: Utilizes a two-inch larger diameter wheel in combination with a tyre that has a two-step lower aspect ratio than the OE tyre.

These days there are numerous choices available for buyers when it comes to cars and almost all car model comes with a variety of choices for the customers which are basically divided into 3 categories. They are known as the base model or the lowest cost model which basically features minimum facilities and hence the tyre given in that model is of the smallest size.

Hyundai i20 Model Car		
Base Model	Medium model	Top end model
Era	Sports+	Asta (o)
Zero	Plus One	Plus Two
185/70 R 14	185/65 R 15	195/55 R 16
SW 129.5mm	SW 120.25	SW 107.25
Rim355.60 mm	Rim381mm	Rim 406.40mm
TH : 614.6mm	TH : 621.5mm	TH : 620.9mm

Where 'SW' stands for Sidewall of the tyre and 'TH' stands for Total Height of the tyre. Thus, in this car, we see that even with plus one upsizing or plus two upsizing the total height of the tyre more or less remains the same and hence all the OEMs today follow this policy so that a buyer gets many options to choose from depending upon his budget and personal preferences.

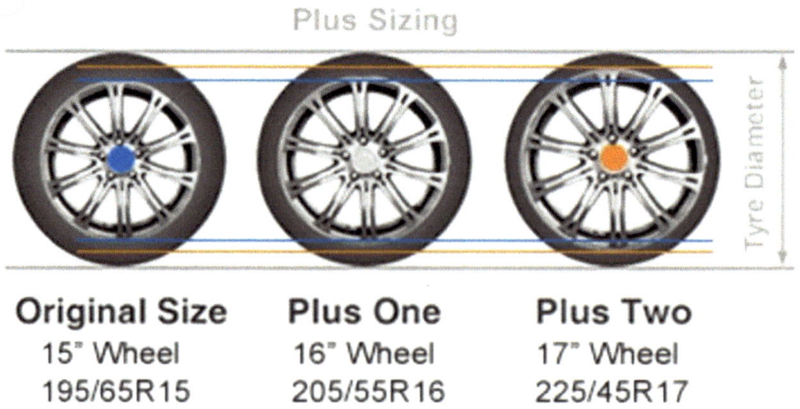

This is how upsizing works just by changing the aspect ratio of the tyre, that is the sidewall height.

CHAPTER 7

INDIAN TYRE INDUSTRY AND MAJOR BRANDS

The Indian tyre industry is an integral part of the Auto sector and it contributes 3% of manufacturing GDP and 0.5% of the total GDP directly. It stands at INR 60-65000 crore industry per annum and 90% approximately comes from the domestic market.

The main players in the Indian market are MRF, Apollo, JKTyre, Ceat and BKT apart from some brands specializing in two / three wheelers' tyres like TVS Eurogrip and Ralco.

Michelin, Bridgestone, Continental, Goodyear are other major international players in the PCR [Passenger Car Radial] and Truck segment that are present in India.

Tyres are generally either supplied to the OEM [Original Equipment Manufacturer] or provided to the replacement market apart from exports.

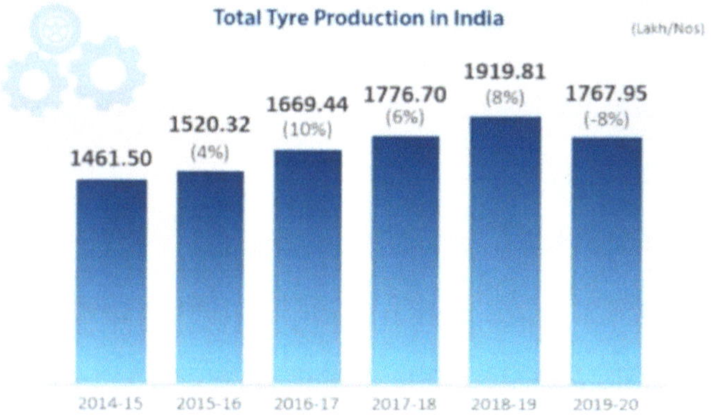

Data Credit: Equirus securities

The share of the replacement market is approximately 70% as compared to the OEM as the amount fetched by tyre companies are much higher in the replacement market sold via Brand retail stores or Multi dealer shops.

Exports from India are to the tune of 10% or lesser.

T&B (Truck and Bus) tyres contribute nearly 55% of revenue in the Indian tyre industry whereas, in the Global tyre industry, it's the PCR that contributes to the largest portion of the revenues.

It is because the penetration of cars per 1000 people is quite low in India, approximately 70 per 1000 whereas it is 250 vehicles per 1000 people in China and almost a car per person in most developed countries with the USA leading the way.

Top Indian Tyre Companies: 2020

MRF: Madras Rubber Factory [MRF] is one of the oldest and largest tyre manufacturers in India. It was set up like a toy balloon manufacturing company in Madras in 1946 by KM Mappiilai and not until 1952 when it turned into tread rubber manufacturing. It is the leading manufacturer of tyres in India. Has got a presence in 65 countries today.

- Total sales: Rs 16711Cr
- Profit: Rs 1087Cr
- Market Cap: Rs 26751Cr
- www.mrftyres.com

Apollo: Apollo tyres limited came into inception in 1972. With corporate headquarters in Gurgaon, it caters to over 100 countries with a manufacturing base in Holland and Hungary apart from India. Led by a dynamic leader in Neeraj Kanwar, has for the first time left MRF behind to become India's leader in tyre sales in 2020.

- Total sales: Rs 17309Cr
- Profit: Rs 507Cr
- Market Cap: Rs 9633Cr
- www.apollotyres.com

MRF VS APOLLO AND THE REST : HOW THEY STACK UP					
COMPANY	**REVENUE**				
	2015-16	**2016-17**	**2017-18**	**2018-19**	**2019-20**
MRF	20,244	14,743	15,227	15,837	16,711
APOLLO	8,702	9,924	10,388	12,354	17,309
JK	5,880	6,608	6,579	7,613	10,167
CEAT	5,494	6,377	6,330	6,831	6,954

- JK Tyre:

  JK Tyre & Industries Ltd. is the 3rd largest tyre maker and among the top 25 manufacturers in the world. Right from 2-wheeler tyres to 13" tall Off-The-Road tyres, it manufactures tyres of all sizes in its manufacturing plants in India and Mexico. Dr Raghupati Singhania is the chairman of the JK group with its headquarters in New Delhi. It has got a strong dealer network partners of over 4000 dealers and 500 Steel Wheels and Xpress Wheels brand stores.

- Total Sales: Rs 10167Cr
- Profit: Rs 134Cr
- Market Cap: Rs 1709Cr
- www.jktyre.com

Indian Tyre Industry, 2020: At a glimpse

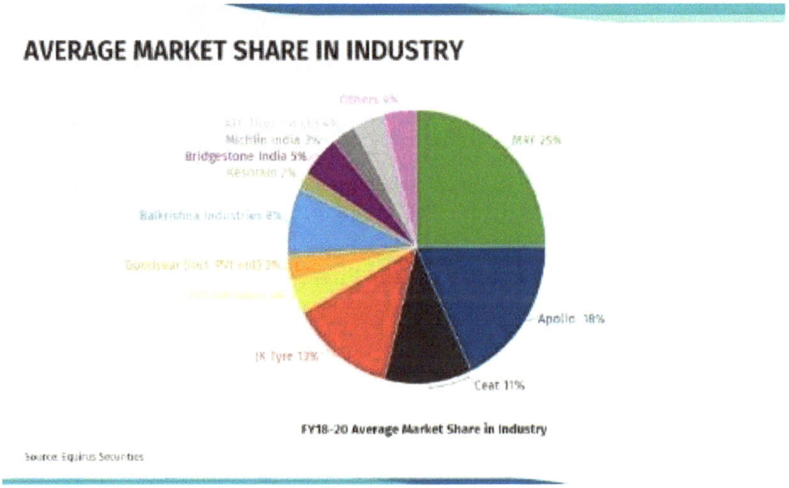

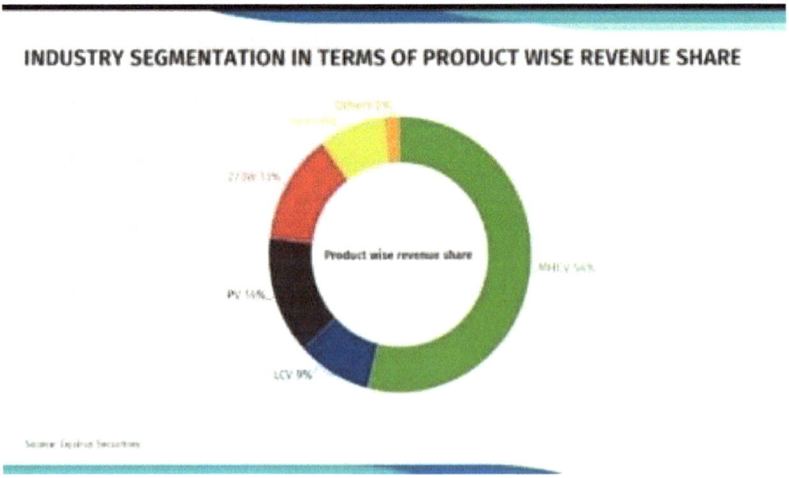

(Usage of Radials in various segment)

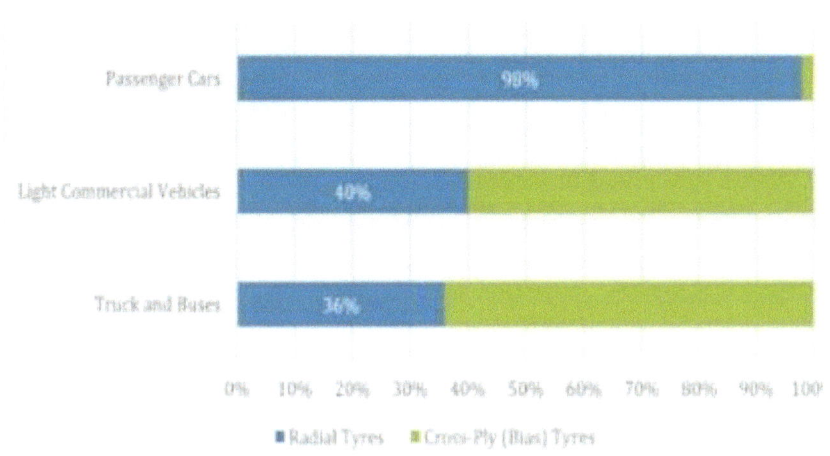

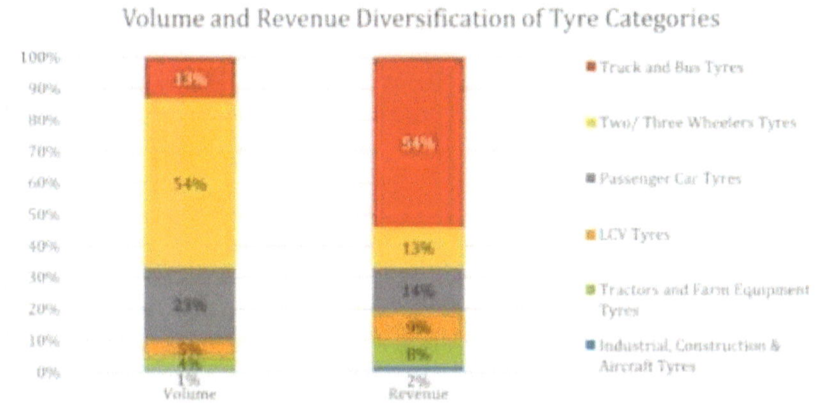

Source: ATMA, Industry and TARI Estimates

- Ceat

  Mumbai headquartered Ceat, the flagship company of RPG was established in 1958. It again produces tyres for all segments with the highest market share in 2-wheeler segment.

- Total Sales: Rs 6954Cr
- Profit: Rs 244Cr
- Market Cap: Rs 3805 Cr
- www.ceat.com

TYRE PLANTS IN INDIA

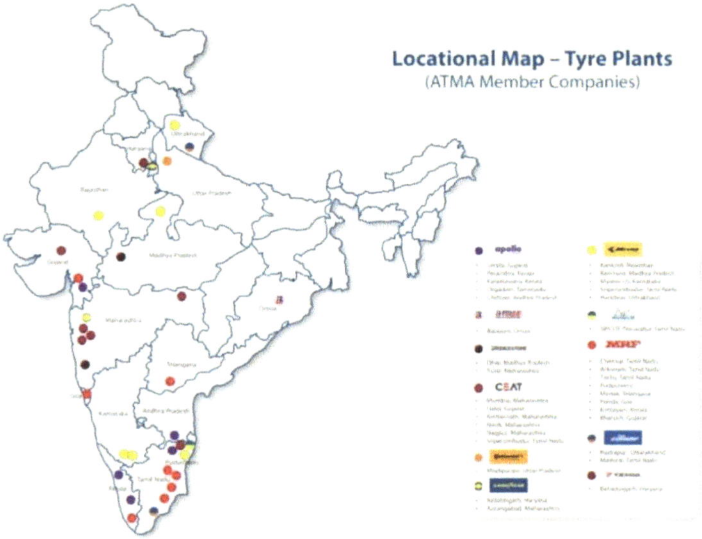

- BKT:

  Bal Krishna Industries Ltd. Or BKT is a leading manufacturer of OTR {Off-The-Road} tyres. Founded in 1987 with its head office in Mumbai. It enjoys 6% market share.

- Total Sales: Rs 4838Cr
- Profits: Rs 809Cr
- Market Cap: Rs 17089Cr
- www.bkt-tires.com

- TVS Eurogrip:

  TVS Srichakra Ltd is one of the leading 2-wheeler tyre manufacturers. Has presently named its brand as TVS Eurogrip. Founded by T V Sundaram Iyengar, is a part of TVS Group which is the 3rd largest 2-wheeler automobile manufacturer also.

- Total Sales: Rs 2354Cr
- Profits: Rs 103Cr
- Market Caps: Rs 1349Cr
- www.tvseurogrip.com

Market share in replacement tyre market in India: 2020 Estimates only						
Company	Truck	LCV	PCR	FARM	2W	OTR
MRF	29	36	15	32	34	21
Apollo	28	30	17	16	3	19
JKTyre	19	19	15	10	10	20
Ceat	11	12	11	9	23	5
TVS					19	
Bridgestone	5		21			2
Goodyear			10	24		
Michelin	3		3		3	2
Continental	2		3			
BKT	0	0		6		22
Others	3	3	5	3	8	9
Total	100	100	100	100	100	100

ATMA or Automobile Tyre Manufacturers' Association is a group formed way back in 1975, registered under the Companies Act as the representative body of the tyre industry of India. The primary function of the Association is to be a conduit between the government and the tyre companies in having two-way communication. 9 tyre companies, who are a member of the association control over 90% of total tyre production in the country.

The main purpose of ATMA are:

- Act as a bridge between the government and tyre manufactures of India
- To address the concerns of the industry by discussing with the government on a regular basis
- For better interaction with the suppliers of natural rubber and other raw materials
- To take the Indian tyre industry to the world
- To finetune the Indian tyre industry with the best technology in the world.
- Present Chairman is Mr KM Mammem, CMD of MRF Ltd., and Anshuman Singhania, the CMD of JK Tyre is the new Vice Chairman appointed recently.
- The customers for the tyre industry are the vehicle manufacturers [OEM] and the suppliers are the producer of natural rubber who come together

along with government officials to discuss matters to take the industry further smoothly.

- ATMA members are MRF, Apollo, JKTyre, Ceat, TVS Eurogrip, Michelin, Bridgestone, Goodyear and Continental.
- www.atmaindia.org.in
- atma@atmaindia.org.in

Tyre and Environment by ATMA:

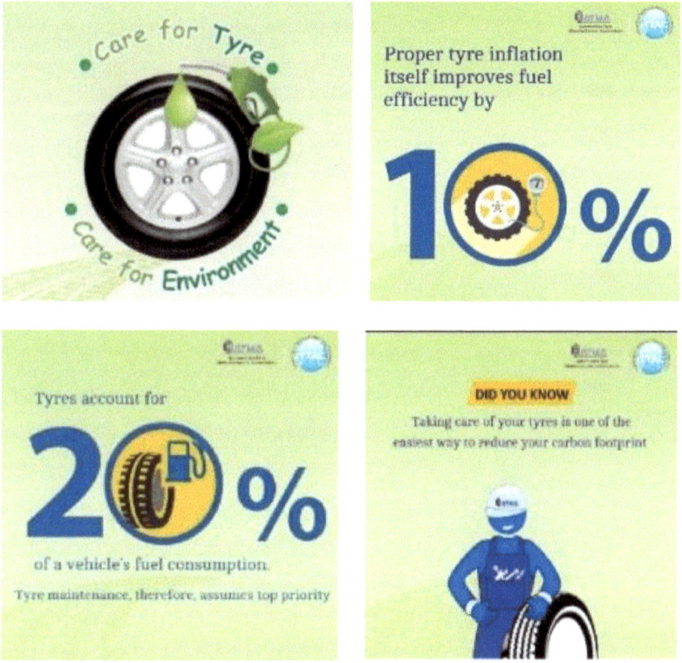

List of All Tyre Manufacturing Companies:

Tire manufacturers

Company	Country	Found.	Brands
Alliance Tire Company	Israel	1950	Alliance[1]
Apollo Tyres	India	1972	Apollo,[2] Apollo Tyres South Africa[a], Kaizen, Maloya, Regal, Vredestein
Birla Tyres	India	1991	Birla Tyre[3]
Belshina	Belarus	1965	Belshina[4]
Bridgestone	Japan	1931	Bridgestone,[5][6][7] Firestone, Fuzion, Seiberling (Latin America), Uniroyal (Australia), Dayton, Supercat (Australia and New Zealand), Fireforce (South Africa)
Bridges Sabanci[8]	Turkey	1974	Bridgestone (Europe and Middle East), Lassa
Carlisle	USA	1917	Carlisle Transportation Products[9][b]
Casumina	Vietnam	1976	Casumina,[11] Euromina
CEAT Ltd	India	1958	CEAT[12]
Cheng Shin Rubber	Taiwan	1967	Cheng Shin, Maxxis,[13] CST,Presa
Continental AG	Germany	1871	Continental,[14][15] Barum, Euzkadi, General Tire, Gislaved, Mabor, Matador, Minerva(Belgium), Point S,[16] Semperit, Sime Tyres, Viking, Uniroyal (Europe), Sportiva, Ameri*Steel, ContiRe, ContiTread, Hoosier Racing Tire
Cooper Tires[17][18]	USA	1914	Avon, Cooper, Dean, Eldorado, Mastercraft, Mentor[c], Starfire, Definity, Roadmaster, Mickey Thompson, Dick Cepek, Chengshan, Austone, Fortune, Hercules[d], Ironman
Deestone	Thailand	1977	Deestone,[19] Thunderer
DMACK	UK	2008	DMACK[20]

Federal Corporation	Taiwan	1954	Federal,[21] Hero, Atturo
General Tyre Pakistan	Pakistan	1963	General[22]
Giti[23]	Singapore	1951	Giti Tire,[24] GT Radial,[25] Primewell,[26] Runway, Dextero, Hualin, Greatwall, Roadpro
Goodyear[27]	USA	1898	Dębica (Poland),[28] Douglas, Dunlop (North America, Europe, Australia and New Zealand), Fierce, Fulda, Goodyear, Kelly, Sava (Slovenia), Pneumant Tires (Pneumant Reifen GmbH)[29]
Hangzhou Zhongce	China	1958	Chaoyang, CYT, Goodride, Westlake, Yartu
Husain Tire[30]			
Hankook Tire[31][32][33]	South Korea	1941	Aurora, Hankook, Kingstar, Laufenn
Hoosier Racing Tire[34]	USA	1957	Hoosier
Hutchinson SNC[35]	France	1957	Hutchinson Tires
Inoue Rubber Co., Ltd.[36]	Japan	1926	IRC Tires
J. K. Organisation[37]	India	1888	JK Tyres, Tornel
Kenda Rubber[38]	Taiwan	1962	Kenda, Kenda radial
Kesoram[39]	India	1919	Birla, KRM, Springfield, AFFINITY
Kumho Tires[40][41]	South Korea	1960	Admiral, Marshal, Kumho, Zetum, Trailfinder[42]
Madras Rubber Factory[43]	India	1946	MRF Tyres
Michelin Group[44]	France	1889	BFGoodrich,[45] Kleber,[46] Kormoran (Poland), Michelin, Orium, Riken,[47] Taurus (Hungary), Uniroyal (North America),

			Tigar Tyres, Strial, Sebring, Stomil Olsztyn (Poland), Camso,[48][49] Solideal[50]
Multistrada Arah Sarana[e][51]	Indonesia	1988	Achilles, Corsa, Strada
Nankang Rubber[52][53]	Taiwan	1959	Nankang, Sonar, Star Performer TNG
Nexen Tire[54][55][56]	South Korea	1942	Nexen, Roadstone, Capitol (joint venture), Lexani
Nokian Tyres[57]	Finland	1898	Nokian, Nordman
Omni United[58]	Singapore	2003	Radar Tires, Goodride(tires), Corsa (tires), Achilles (tires), RoadLux (tires), Patriot (tires) (https://myjeeppatriot.com/)
Panasonic	Japan	1953	Panaracer
Petlas	Turkey	1976	
Pirelli[59][60]	Italy	1872	Agom, Courier, Metzeler, Pirelli, Formula[61]
Shandong Linglong Tire Co., Ltd.[62]	China	1975	Linglong, GREEN MAX, LEAO, ATLAS, INFINITY
Servis Tyres	Pakistan	1970	Servis
Shanghai Huayi[63]	China	1996[64]	Double Coin (Warrior)
Starco[65]	Denmark	1962	Tusker
Stomil-Poznań[66]	Poland	1928	Stomil[f]
Sumitomo Rubber[67][68]	Japan	1909	Blacklion Tires (USA), Dunlop (Japan, Latin America, Asia and Africa), Falken,[69] Goodyear (Japan), Sumitomo, Multi-Mile, Sigma, Cordovan, Vanderbilt, Ohtsu (https://ohtsu-tires.com) , Solar

Suryaraya Rubberindo Industries	Indonesia	1991	FDR
Titan Tire Corporation[70]	USA	1993	Continental (farm tires), Goodyear (farm tires), Titan
Toyo Tire & Rubber[71]	Japan	1945	Nitto, Silverstone, Toyo
Trelleborg AB[72]	Sweden	1905	Trelleborg, Cultor Tyres (Europe), Mitas (Czech Republic), Rubena (Czech Republic)
Triangle Group[73]	China	1976	Triangle Group, Diamond back
TVS Srichakra Ltd.[74]	India	1982	TVS Tyres, TVS Eurogrip[75]
Viking[76]	Norway	1931	
Vittoria	Italy	1953	Vittoria, Geax
Vogue Tyre and Rubber	USA	1914	
Yokohama Rubber[77]	Japan	1917	Yokohama, Advan
Zenises[78]	UK	2014	Z Tyres, T Tyres, Westlake, Triangle, iLink, Kapsen, Ardent
Elangperdana Tyre Industry, PT	Indonesia	1996	ACCELERA, FORCEUM, ZEETEX

Notes

a. *formerly Dunlop South Africa*

b. *Lawn and garden, ATV, golf car, trailer, skid steer, agriculture, industrial, styled steel wheels and related replacement markets.*[10]

c. *brand introduced in spring 2012*

d. *Cooper is the contract manufacturer*

Source: Wikipedia

CHAPTER 8

GLOBAL TYRE INDUSTRY

The global tyre industry size was valued at $167US billion in 2019. The world tyre industry is led by Michelin, Bridgestone, Goodyear among others. The Indian brands to feature in the top 25 companies are MRF, Apollo, JKTyre and Ceat. Worldwide, Bridgestone itself sells $25 US billion and so does Michelin. Thus, the Indian tyre market in comparison to the world market is quite small taking its size and population into consideration which is only $8.5 US billion which means, almost, 5% market share of the global tyre industry.

Truck & Bus (T&B), Passenger Vehicle (PV), 2/3-Wheeler, Off-Highway Tyres (OHT) & Others

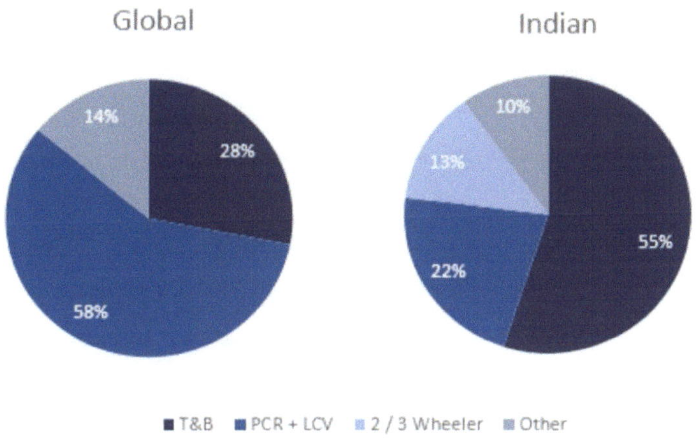

Data Source: Alphainvesco.com

World's top automobile manufacturer and thereby tyre consumer:

As we know, tyres are directly related to automobile manufacturing and usage. The countries that manufacture maximum automobile needs maximum tyres and the top automobile using countries needs tyre in replacement markets if not as OEM.

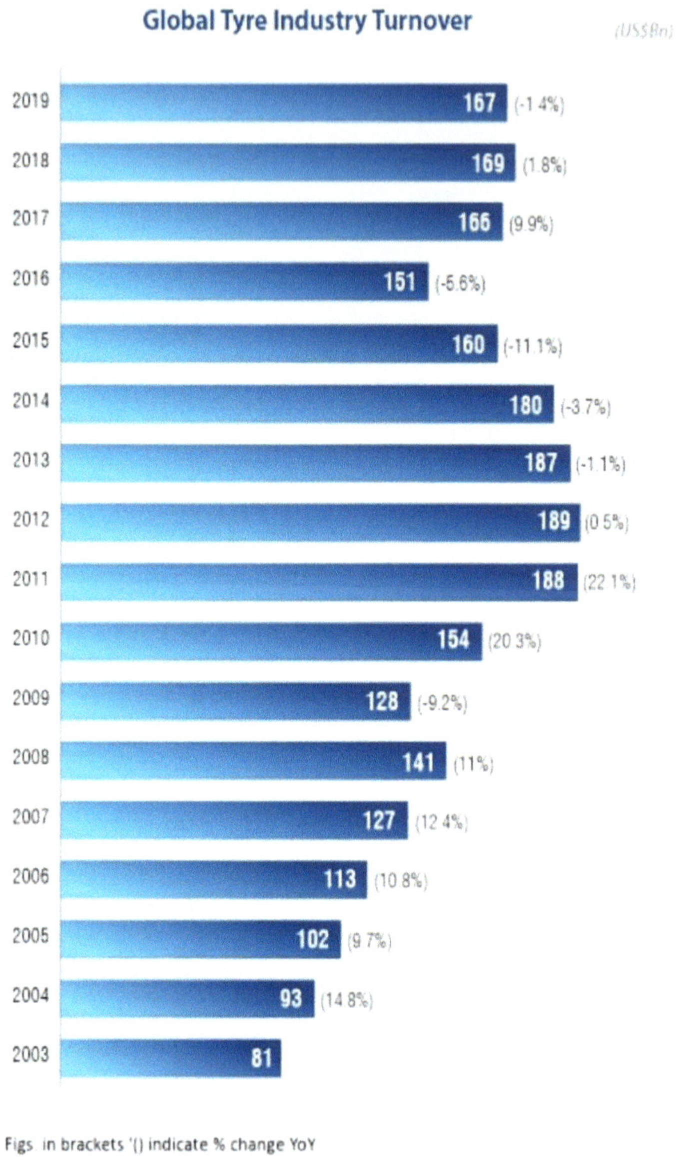

Data Source: ATMA

TOP AUTOMOBILE MANUFACTURERS OF 2020

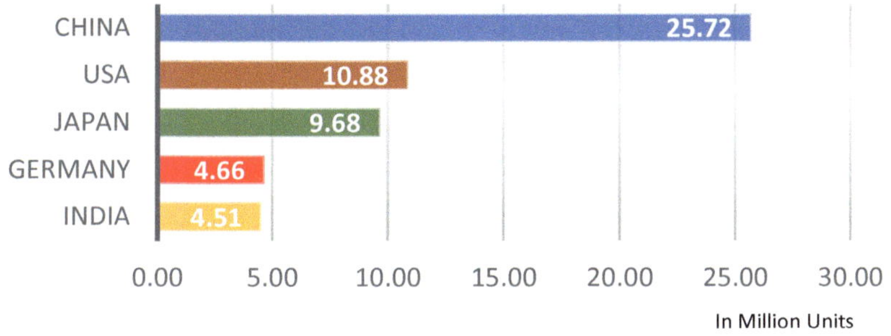

TOP AUTOMOBILE MANUFACTURERS OF 2020:		
		(In million units)
1	China	25.72
2	USA	10.88
3	Japan	9.68
4	Germany	4.66
5	India	4.51

The OEM are huge customers of the tyre industry which comprise nearly 30%, exports to the tune of 10% and replacement markets make the rest 60%. Also, the value realised is much better from the replacement market.

Top 10 Rubber Producing Countries in 2020		
Rank	**Country**	**Avg rubber Production per Annum (1000 Tons)**
1	Thailand	4305
2	Indonesia	3088
3	Malaysia	997
4	India	891
5	China	864
6	Vietnam	790
7	Philippines	548
8	Cote d'Ivoire	411
9	Guatemala	356
10	Brazil	186

Leading Consumers of Natural rubber Worldwide				
				(In 1000 Metric Tons)
Characteristic	**2014**	**2016**	**2018**	**2019***
China	4760	4982	5504	5497
India	1015	1033	1220	1144
United States	932	932	987	1006
Thailand	541	650	752	800
Japan	709	676	706	714
Indonesia	540	583	618	625
Malaysia	447	486	515	501
Brazil		412	398	400
South Korea	402	381	367	354

Analysis:

We have seen the leading automobile manufacturing countries followed by leading natural rubber producing countries and finally, the leading consumers of natural rubber.

China, as we saw, is by far the leading automobile manufacturer and is not a large producer of rubber but again is the leading consumer. Similar, is the case for India and the United States and hence these countries have to rely on imports of natural rubber and their price competitiveness relies on the import duties and other governmental levies. Indian tyre manufacturers major concern today is the lack of sufficient natural rubber in the country and they need to rely on imports which attract 25% duties as against finished tyre which attracts only 15% import duties.

Analysing the above figures shows that China would be importing around 80% of natural rubber followed by the US, approximately, 80% again and India too depends on other countries to the extent of 50%.

Thailand, Indonesia and Malaysia on the other hand are the countries where the production of natural rubber is much higher than consumption and hence are the largest exporters of natural rubber.

WORLD'S TOP TYRE MANUFACTURING COMPANIES:

1.. MICHELIN

Michelin is a French multinational tyre manufacturing company and the largest tyre manufacturer before Bridgestone took over with revenues of $25.47 US billion. It was formed in 1889 when two brothers Edouard Michelin and Andre Michelin came together. The Michelin Man, Bibendum, is a world-famous and more than a hundred-year-old mascot.

- www.Michelin.in
- www.tyres-n-services.in

2. BRIDGESTONE

Bridgestone is a Japanese multinational tyre manufacturing company founded in 1931. Bridgestone group had 181 production facilities in 24 countries as of July 2018. The sales are approximately $25 US billion.

They are doing very well in the passenger car market in India with a 25% market share.

- www.bridgestone.co.in

3.

The Goodyear tyre and rubber company is an American multinational tyre manufacturing company founded in 1898 by Frank Seiberling in Ohio, USA. Dunlop tyres and Cooper tyres are subsidiaries of Goodyear tyre company.

- www.goodyear.co.in

4.

Continental AG tyre is a German multinational company and is the world's fourth largest tyre manufacturer founded in 1871 with its headquarters in Hanover, Germany. It is one of the pioneers in tyre manufacturing with many patents to its name in tyre evolution and has just clocked in 150 years of existence.

- www.continental-tyres.in

5.

Pirelli & C.S.p.A. is a multinational tyre manufacturer based in Milan, Italy. It has 19 manufacturing sites in 13 countries and a network of 14600 distributors and retailers. Pirelli was founded in 1872 by Giovanni B Pirelli.

- www.pirelli.com

6\.

Sumitomo rubber industries limited is a global tyre and rubber company based in Japan. Sumitomo brands include Falken tyres, Dunlop tyres and Ohtsu tyres. It was founded in 1909. But the Falken tyre brand was launched by Sumitomo in Japan in 1983. And the Dunlop tyre brand is partly owned by Sumitomo. In India the brand is operated by Dunlop India limited.

- www.falkentyre.in

7\.

Hankook tyre is a South Korean tyre company, based in Seoul. It is the 7th largest tyre company in the world and was founded in 1941. It was founded in the name of Chosun tyre company and was later renamed to Hankook in the year 1968.

- www.hankooktire.com

8\.

ZHONGCE RUBBER GROUP CO., LTD.

It is China's largest manufacturer of automotive tyres and features in the top 10 tyre makers of the world with sales of $4.26 US billion. It was founded in 1958.

- www.zc-rubber.com

9. YOKOHAMA

The Yokohama rubber company limited is a tyre company based in Tokyo, Japan. The company was founded in 1917 and features in the world's top 10 tyre manufacturers.

- www.yokohama-india.com

10. CHENG SHIN TIRE

Established in 1967 in Taiwan, Maxxis tyres and CST tyres are a wholly-owned subsidiary of Cheng Shin. It was founded by Loujie, which began as a producer of bicycle tyres and expanded into other types with a revenue of over $3.85 US billion.

- www.csttires.com

World's Leading Tyre Makers:

The world's leading tyre manufacturers

Updated: 3 July 2021

Rank	Company	Note	2020 Turnover Euro	2020 Proportion tyres	2020 Proportion tyres (%)	2020 Local curr.	2019 Euro	2019 Local curr.	2018 Euro	2018 Local curr.
1. (2.)	Michelin		20,469.0	20,059.6	98.0%		24,135.0		22,028.0	
2. (1.)	Bridgestone	1)	23,673.8	19,649.3	83.0%	2,994.5	28,912.6	3,525.6	29,003.7	3,650.1
3. (4.)	Continental		37,722.3	10,158.6	26.9%		44,478.4		44,404.4	
4. (3.)	Goodyear	2)	10,040.7	10,040.7	100.0%	12,321.0	13,125.3	14,745.0	13,515.3	15,475.0
5. (5.)	Sumitomo Rubber Industries	1)	6,251.9	5,376.6	86.0%	790.8	7,325.8	893.3	7,105.6	894.2
6. (7.)	Hankook	3)	4,830.8	4,830.8	100.0%	6,454.0	5,310.0	6,883.3	5,317.3	6,795.1
7. (6.)	Pirelli		4,302.0	4,302.0	100.0%		5,323.0		5,194.5	
8. (8.)	Yokohama	1)	4,511.0	3,775.7	83.7%	570.6	5,334.3	650.5	5,166.8	650.2
9. (9.)	zc rubber	4)	3,508.6	n/a		28,148.0	3,527.7	27,588.0	3,413.5	26,881.9
10. (10.)	Maxxis	5)	2,790.4	2,790.4	100.0%	96,209.0	3,254.2	109,508.0	3,120.2	109,221.0
11. (12.)	Toyo Tires	1)	2,718.0	2,427.2	89.3%	343.8	3,095.4	377.5	3,124.5	393.2
12. (14.)	Linglong Tire	4)	2,291.4	n/a		18,382.7	2,194.8	17,164.2	1,943.1	15,302.0
13. (11.)	Giti	10)	2,179.0	n/a			2,760.0		2,807.9	
14. (13.)	Cooper Tires	2)	2,054.5	2,054.5	100.0%	2,521.1	2,450.2	2,752.6	2,452.5	2,808.1
15. (15.)	apollo	6, 9)	2,042.4	n/a		175,263.8	1,969.5	163,270.0	2,258.0	175,488.4
16. (17.)	Sailun Tires	4)	1,920.1	1,920.1	100.0%	15,404.0	1,934.4	15,127.8	1,737.8	13,685.0
17. (16.)	MRF Tyres	6, 9)	1,883.5	n/a		161,631.9	1,958.9	162,393.6	2,066.7	160,624.6
18. (18.)	Kumho Tyre	3)	1,624.8	1,600.4	98.5%	2,170.7	1,827.7	2,369.2	2,002.2	2,558.7
19. (19.)	Nokian Tyres		1,313.8	1,313.8	100.0%		1,595.8		1,595.6	
20. (20.)	Nexen Tire	3)	1,271.0	n/a		1,698.1	1,560.1	2,022.3	1,552.5	1,984.0
	Huayi / Double Coin	4)	3,506.0	1,051.8	~ 30%	28,126.8	4,798.4	37,526.1	5,617.7	44,240.0
	JK Tyre	6, 9)	1,065.7	n/a		91,452.7	1,055.9	87,532.9	1,310.9	104,520.1
	Triangle	4)	1,063.9	1,063.9	100.0%	8,535.3	1,015.4	7,940.8	953.8	7,511.0
	Titan	2)	1,026.2	1,026.2	100.0%	1,259.3	1,289.5	1,448.7	1,399.5	1,602.4
	CEAT	6, 9)	888.4	n/a		76,234.1	798.9	66,224.5	886.1	68,866.0
	Kenda	5)	877.0	877.0	100.0%	30,260.2	954.7	32,127.4	900.7	31,526.0
	Trelleborg	8)	3,272.4	873.5	26.7%	32,836.0	3,502.3	36,588.0	3,316.0	34,005.0
	Advance Tyre (Guizhou Tyre)	4)	848.7	n/a		6,808.7	825.8	6,458.3	866.6	6,825.0
	Prinx Chengshan	4)	783.2	n/a		6,283.1	714.7	5,589.0	661.1	5,206.0
	Aeolus	4)	695.4	n/a		5,578.7	756.2	5,914.2	789.7	6,219.0
	BKT	6, 9, 10)	694.0	694.0	100.0%	59,553.5	610.6	50,620.2	698.4	54,281.6
	Doublestar	4)	551.5	n/a		4,420.7	527.3	4,123.8	475.6	3,745.0
	Brisa	7)	464.9	464.9	100.0%	4,236.9	532.3	3,558.3	495.0	2,998.8
	Nankang	5)	281.2	n/a		9,695.1	330.2	11,111.6	298.4	10,445.3

1) billion Yen 2) million US-Dollar 3) billion Korean Won 4) million Chinese Yuan 5) million Taiwan-Dollar 6) million Indian Rupee 7) million Turkish Lira 8) million Swedish Krona
9) Financial year April to March 10) Numbers supplied on request

The world's leading tyre manufacturers

Updated: 10 July 2020

Rank	Company	Sales 2019 Euro	2019 Share from tyres		2019 Local curr	2018 Euro	2018 Local curr	2017 Euro	2017 Local curr
1. (1.)	Bridgestone [1]	28,912.6	24,217.7	83.8%	3,525.6	29,003.7	3,650.1	26,986.2	3,643.4
2. (2.)	Michelin	24,135.0	23,652.3	98.0%		22,028.0		21,960.0	
3. (3.)	Goodyear [2]	13,125.3	13,125.3	100.0%	14,745.0	13,515.3	15,475.0	12,821.6	15,377.0
4. (4.)	Continental	44,478.4	11,728.0	26.4%		44,404.4		44,009.5	
5. (5.)	Sumitomo Rubber Industries [1]	7,325.8	6,294.5	85.2%	893.3	7,105.6	894.2	6,502.2	877.9
6. (7.)	Pirelli	5,323.0	5,323.0	100.0%		5,194.5		5,352.3	
7. (6.)	Hankook [3]	5,310.0	5,310.0	100.0%	6,883.3	5,317.3	6,795.1	5,324.6	6,812.9
8. (8.)	Yokohama [1]	5,334.3	4,293.0	80.5%	650.5	5,166.8	650.2	4,786.9	646.3
9. (9.)	ZC Rubber (Hangzhou Zhongce) [7]	3,527.7	n/a		27,588.0	3,413.5	26,881.9	3,063.4	24,473.4
10. (10.)	Maxxis	3,254.2	3,254.2	100.0%	109,508.0	3,120.2	109,221.0	3,155.9	112,309.0
11. (11.)	Giti	2,760.0	n/a			2,807.9		2,989.4	
12. (12.)	Toyo Tires [1]	3,095.4	2,729.5	88.2%	377.5	3,124.5	393.2	2,999.8	405.0
13. (13.)	Cooper Tires [2]	2,450.2	2,450.2	100.0%	2,752.6	2,452.5	2,808.1	2,380.6	2,855.0
14. (17.)	Linglong Tire [7]	2,194.8	n/a		17,164.2	1,943.1	15,302.0	1,783.4	13,918.0
15. (14.)	Apollo Tyres [4,9]	1,969.5	n/a		163,270.0	2,258.0	175,488.4	1,880.3	150,977.4
16. (15.)	MRF Tyres [4,9]	1,958.9	n/a		162,393.6	2,066.7	160,624.6	1,919.3	154,109.9
17. (186.)	Sailun Tires [7]	1,934.4	1,934.4	100.0%	15,127.8	1,737.8	13,685.0	1,769.1	13,807.0
18. (16.)	Kumho Tyre [3]	1,827.7	1,802.1	98.5%	2,369.2	2,002.2	2,558.7	2,247.9	2,876.4
19. (20.)	Nokian Tyres	1,595.8	1,595.8	100.0%		1,595.6		1,572.5	
20. (-)	Nexen Tire [3]	1,560.1	n/a		2,022.3	1,552.5	1,984.0	1,547.9	1,964.8
- (19.)	Huayi / Double Coin [7]	4,798.4	1,439.5	~ 30%	37,526.1	5,617.7	44,240.0	5,580.6	43,553.0
	Titan [2]	1,289.5	1,289.5	100.0%	1,448.7	1,399.5	1,602.4	1,224.8	1,468.9
	JK Tyre [4,9]	1,055.9	n/a		87,532.9	1,310.9	104,520.1	1,064.0	85,434.1
	Triangle [7]	1,015.4		100.0%	7,940.8	953.8	7,511.0	1,014.9	7,921.0
	Kenda [8]	954.7	954.7	100.0%	32,127.4	900.7	31,526.0	858.0	30,533.0
	Trelleborg [6]	3,502.3	921.6	26.3%	36,588.0	3,316.0	34,005.0	3,208.2	31,581.0
	Tyre Advance (Guizhou Tyre) [7]	825.8	n/a		6,458.3	866.6	6,825.0	891.8	6,960.0
	CEAT [4,9]	798.9	n/a		66,224.5	886.1	68,866.0	800.7	64,291.4
	Aeolus [7]	756.2	n/a		5,914.2	789.7	6,219.0	935.9	7,304.0
	Prinx Chengshan [7]	714.7	n/a		5,589.0	661.1	5,206.0	620.2	4,840.0
	BKT [4,9]	610.6	610.6	100.0%	50,620.2	698.4	54,281.6	597.9	48,007
	Brisa [5]	532.3	532.3	100.0%	3,558.3	495.0	2,998.8	504.6	2,794.1
	Doublestar [7]	527.3	n/a		4,123.8	475.6	3,745.0	512.3	3,998.0
	Nankang [8]	330.2	n/a		11,111.6	298.4	10,445.3	318.3	11,326.0

1) billions Yen 2) millions US Dollars 3) billions Won 4) millions Indian Rupees 5) million Turkish Lira 6) millions Swedish Krona 7) millions Chinese yuan 8) millions Taiwan Dollars
9) Reporting from April to March

Data Credit: tyrepress.com

World's Largest and Costliest Tyres

World's-Largest-Tire

The world's largest tyre is located in Allen Park, Michigan, USA. It weighs 12 tons and is 80 feet tall.

#1 Gold Plated Z Tyres

Gold plated and diamond studded tyres is developed by Z company of Dubai. Set of 4 tyres will cost you a cool INR 4 Crore or $US 600000.

#2 Letourneau L-2350 Front-loader Tyres

World's largest tyre is 14 feet in height and weighs 6 tones. It cost is INR 42.21 lacs per piece.

#3 Caterpillar 797 Dump Truck Tyres

These tyres are made for caterpillar dump trucks and each costs INR 28.5 lacs

#4 Bugatti Veyron Tyres

Made by Michelin for fastest car to run at 400kmph, it costs approx. INR 26.8 lacs

#5 Space Shuttle Tyres

Each aircraft tyre costs INR 3.7 lacs and is good for single landing only.

Michelin's Largest Tyre.

Largest Tyre Manufacturer.

India's Largest Tyre.

CHAPTER 9

TYRE BRAND STORES

Tyre Buying is an important experience that one should experience best in a tyre branded store. Almost all top tyre brands that exists in the market have their concept stores at major Indian cities. Not less than 4000 branded stores are present in the country. In India, 50% of tyre buying happens at brand stores as they have expertise in sales and various services too associated with new tyres. The reasons to buy tyres at branded stores are:

- Expertise in tyre sales
- Latest tyres
- Variety of tyres
- Complimentary services like alignment, balancing, nitrogen air, automatic fitment among others
- Other associated services like battery checking, pollution testing, AC gas filling, Car detailing, etc.

Brand Stores images of some major Tyre Brands:

CHAPTER 10

TYRE CARE AND MAINTENANCE

"A Car and its tyres are like husband and wife.
Lack of understanding and care leads to trouble!!"

A Tyre, as we have seen earlier, is the only contact area between the vehicle and the road and hence becomes a very integral part of your vehicle.

It bears the weight of the vehicle at all times whether on the run or still and bears the friction on the road, leading to its wear and tear throughout its lifetime.

Some gruesome facts:

Do you know the number of road accidents that take place on an average in a year in India alone?

It is a huge 150000 people which makes for approximately 400 deaths per day.

Year	No. of fatalities in 000's
2019	151.11
2018	151.42
2017	147.91
2016	150.79
2015	146.13
2014	139.67
2013	137.57
2012	138.26

Can you imagine, how many of them are because of tyre-related issues such as tyre bursts and flat tyres?

As much as 20% of all road accidents.

How many of the tyre bursts are because of inflation issues, tyre overheating issues, tyre alignment issues? As much as 70% of vehicle tyres are either overinflated or underinflated but not correctly inflated as per manufacturer recommendations and 50% of the tyres that burst are worn-out tyres which were running even after crossing the tyre wear indicator mark, that asks us to replace our tyres.

Accidents also take place because of overloading and improper loading which puts extreme pressure on tyres resulting in tyre bursts.

Thus, these precious lives can be saved if proper care is taken for the vehicle tyre to a great extent if not all.

Tyre and Shoes: A comparison!

We can compare vehicle's tyres and its tread to our shoes too where we care what to buy and what to consider before buying. We take it to account not only the aesthetics but also the heal and the sole which provides the grip or traction. We very much check and go for it after we are assured of the safety factor and then why not understand a tyre before choosing them for our vehicle.

After all, we can walk and run without shoes but we can't imagine a vehicle moving without tyres.

Hence, we can say that "it all rides on our tyres" and its care and maintenance become that much more important and necessary too.

To understand and know how to maintain your car tyres in the best possible manner, we need to understand the basic roles that a vehicle tyre fulfils.

The major **functions** of a tyre are to provide:

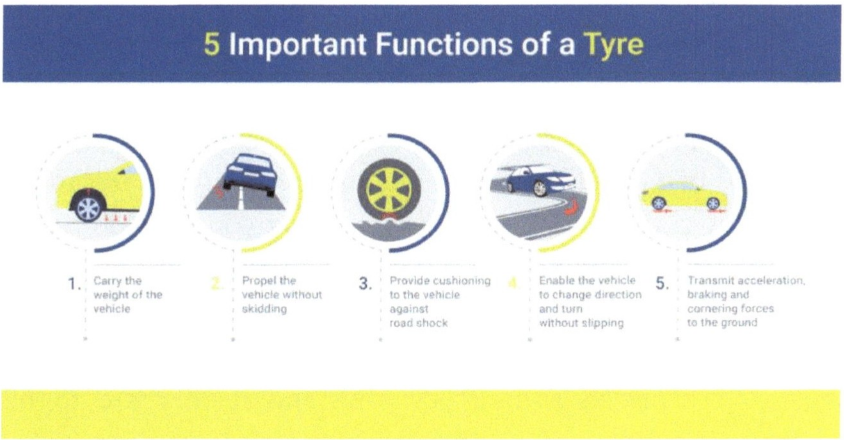

- smoother motion:

  it is the tyre of a vehicle that rotates and takes us from one place to another by transmitting the engine power into acceleration and braking. And the minimum thing we expect from our tyres is a smoother journey to our destination. Hence the tyres need to be in proper shape and thoroughly checked before a journey for a smoother drive.

- comfort:

  the second most important role that I think a tyre should enable is the comfort of the passengers. Since the evolution of tyres over the centuries, we have seen, comfort, has been a major improvement. The more the cushion effect in a tyre, the more comfortable the drive it becomes for the passengers. Hence radialisation of cars has been a major factor in the improvement of comfort factor as the sidewalls also take lots of load along with the tread area. Thus, the tyre as a whole need to be checked and maintained for a comfortable drive.

- safety:

  the safety of all passengers in a car is again of utmost importance where the tyres of a vehicle play a major role as any punctures or a tyre burst because of any negligence could be fatal. Thus, when it comes to safety, there cannot be any compromises and as a customer you should opt for nothing but the best. Cost does matter but quality should be given priority when choosing to buy new tyres and deciding whether a damaged tyre can be repaired or needs to be replaced with a new one and if they are to be replaced, how and in what manner. It's better to ask for expert guidance from a brand store.

- speeding and braking:

  as discussed earlier, the speeding and braking functions are also performed by tyres on the road. Braking is a very important function as the sooner the vehicle stops, the safer it is for the passengers. Delay in braking because of the damaged tread or lack of sufficient tread on the tyre could prove fatal again. Very often we see that Vehicle users do wait for the whole tread to wear out totally which could be very dangerous as there will be a lack of grip while braking and the tyre will skid resulting in an accident. Hence, tyres need to be replaced on time to avoid such circumstances as skidding or hydroplaning (floating on shallow water on roads because of lack of tread grip).

- direction:

  well, it is the steering and the tyres that play a major role in reaching our destination by giving directions to our motion on the roads.

  The tyres take a lot of pressure in motion, especially in turnings and in ghat roads as there is the involvement of alleviation and centrifugal force. The tread and the sidewalls need to be in top conditions to withstand this pressure in high speeds to avoid any risk. Any damages in sidewalls or tread itself cannot be ignored.

  They need to be fixed before a journey to avoid any further delays during the journey and the risks arising out of it.

- carry the load of the vehicle uniformly:

  by this, we mean that the total load of the vehicle should

be carried by all four or all tyres on an equal basis to maintain the stability of the vehicle. In case of a lack of tread on a single tyre, it can produce inequality in the other tyre on the same axle which again can prove fatal. It is recommended that all the 5 tyres should be replaced at once to maintain uniformity. Proper tyre care which includes tyre rotation, wheel alignment, balancing and proper air pressure will ensure equal pressure on all tyres for better load-carrying capacity of the vehicle as a whole.

Loads and Loading practices

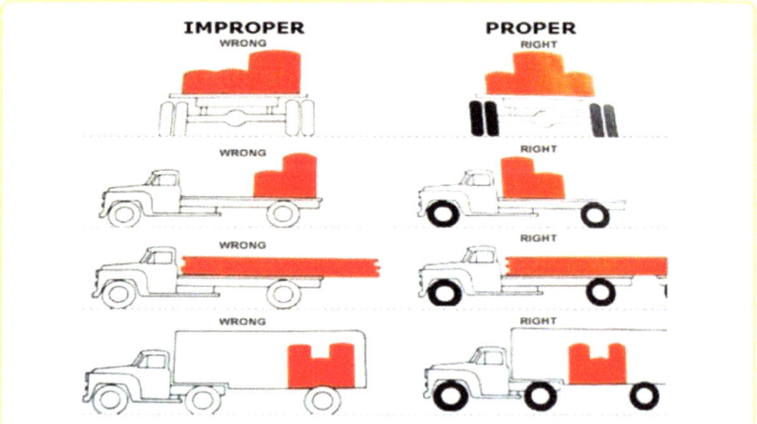

In the above illustration, we see the loading practices for commercial vehicles. The loading should be such as there is even distribution of load on all the tyres equally. The illustration clearly mentions the proper and improper loading. In case of improper loading, there will be an uneven distribution of weight and hence uneven tyre wear. Also, the risk arising out of improper loading could be fatal.

Now let us understand in various ways how we can take care of our car tyres or other vehicle tyres in general for their longevity and smoother drive.

How to take care of your vehicle tyres?

New Car tyres:

Whenever we buy a new car, it obviously comes with sets of 5 brand new tyres from the OEM, 4 fitted and one spare tyre of the same size called as the Stepney tyre.

Make sure to inspect the tyres at the time of taking delivery of the car for any cuts or damages along with Stepney as once delivery is complete and you come to notice of any damage after a couple of days or even a few hours too, it would not be considered by the car showroom as there is every chance of tyre being damaged after the delivery. Hence, an inspection of tyres along with all vehicle features are mandatory too.

Generally, an OEM tyre is of good quality as any remark on the tyre would damage the goodwill of that company; be it Michelin, JKTyre or any other.

As much as, 40 to 50000 kilometres life for a 5-tyre set is said to be fair enough performance if proper tyre care has been taken along the way.

Tyres with new cars or newly replaced tyres carry 2 to 5 years conditional warranty or 20000kms whichever is an earlier clause. Other than manufacturing defects, tyres are not liable for replacements by the authorised tyre dealer or that particular brand store. A lot of tyres come with sidewall cuts or tread cuts which are generally stone hit damages and doesn't come under manufacturing defects.

Tyre Care Tips:

Tyres are meant to wear out eventually. Even the best quality tyres do age with time and need to be replaced whether used or not. However, it is possible to extend the life of the tyres through some basic preventive care and maintenance. They need regular care but before a long journey, tyres need to be checked thoroughly. Here are some basic tips on tyre care you need to follow for better tyre life and a smoother journey.

1. **Air Pressure**:

   It is the air inside the tyre that bears the weight of the vehicle and hence correct air pressure as per the OEM recommendations as mentioned in the owner's manual is of utmost importance.

   Pressure should also be checked before any outdoor or long journey as there is gradual leakage of air even if the vehicle is not moving.

   It is recommended to have an air pressure gauge in your vehicle to crosscheck as required.

- Get your tyres pressure checked at regular intervals
- Have a pressure gauge in your car

- Fill with Nitrogen Air if available:
    - As its lighter than normal air
    - Lasts longer
    - Heat generation is lesser
    - Lesser rusting of steel cords in tyre
    - Better fuel economy
- Get your tyre pressure checked at proper branded stores
- Maintain correct air pressure as per manufacturer recommendations in your car's owner manual.
- **Avoid Over inflation** and **Under Inflation**:

Tyre safety starts with inflation. It is the air inside the tyre that supports the load, grips the road and absorbs the bump. When you drive on the road, each part of the tyre changes its shape numerous times. This causes the flexing of the sidewalls which produces heat. Tyres is designed around the idea of maintaining the flex in the right amount in the right area. The amount of flex depends upon the amount of inflation. For a given load, less inflation means more flex and more inflation means less flex. Both could lead to costly damages.

Too much flex or Under Inflation will lead to wear out of tyre tread near the edges. It also leads to higher heat generation thus increasing fuel consumption too. This increased heat produced causes the bonds of the steel chords to break. Subsequently, if there is too much heat generation, it can lead to total separation which we call tyre burst and this can be lethal as the vehicle can get out of control.

On the other hand, less flex or overinflation leads to more tyre wear at the centre of the tread and a rough and bumpy ride. It also causes many damages and burst when the tyre hits a pothole or any sharp objects at high speed. Thus, correct inflation is the best option for a comfortable and safe drive.

Diffusion or leakage of air in small amounts keeps on happening and hence air pressure should be checked and filled regularly, preferably Nitrogen air.

Thus, avoid Under or Overinflation which will result in risk and damages.

Some of you might have observed, drivers do diffuse air purposefully making them underinflated before driving their vehicle into a desert so that the tyre grip increases in the sand and the safety goes up. But after the completion of the sand dunes drive, they again fill the tyre with correct air pressure for the highway roads. Thus, different terrain or different vehicle load may demand variation in tyre pressure for a shorter duration.

The factors which influence tyre life:

⊞ **Inflation Pressure**

- **Smart Tyres** or **TPMS**:

We are all talking about smart gadgets these days. Be it mobiles, laptops, bikes, cars or others. For your kind information, even tyres can also be converted into smart tyres by implementing TPMS or Tyre Pressure Monitoring System which is an electronic system designed to monitor the air pressure inside the pneumatic tyres on various types of vehicles. A TPMS reports real-time air pressure to the driver either through a pictorial display or a gauge. Smart sensors are fitted into the rim when fitting tyres. Proper action can be taken to make corrections and avoid accidents that could arise due to sudden air leakage. Also, GPS tracking of the vehicle is possible by downloading a smart application on your phone. Though it started in the 1980s in the luxury vehicles, and it stands at 50% usage worldwide; in India, this has been started by JK Tyre company by the name, **Treel.** Let us understand the main purpose behind this:

- Keep an eye on your vehicle even when it is out of sight
- Don't let flat tyres ruin your busy schedule
- Cruise down highways without breakdown

- Cut down on expenses and avoid unforeseen accidents
- Cut down on mileage loss due to underinflated tyres. They drink more fuel per kilometre.
- Ensures maximum performance and safety.

2. Check tyre tread depth:

   When should we replace the tyres with new ones?

   Generally, most vehicle owners assume, it is only after the total tread is over that they should opt for new tyres; but it's not so and it's actually quite dangerous to wait for that long.

   There is something called 'TWI' or **Tread Wear Indicator** mark on your tyre which is 1/16inch of the total tread depth. Once, TWI mark is reached, tyres should no longer be used as there is not much tread left for gripping and braking of the vehicle. The more the tyre tread depth, the better is tyre grip and stopping capabilities. Generally, the tread depth varies from segment to segment.

   For Trucks, it is mostly 20mm plus but for car tyres, it is

8mm in total. So, in-car tyres TWI is marked at 0.5mm tread. Hence, change the tyres as soon as the tread touches the TWI mark.

Total tread wear results into:

- Lack of strong grip on the roads
- Skidding when sudden breaks are applied
- Tyre burst due to reduced strength
- Accidents, as vehicle skids without stopping
- Hydroplaning (floating of tyres) on shallow water cushions on roads

3. Wheel Alignment:

Wheel alignment also referred to as tyre alignment, is the correction of the angles of the tyres so that they come into contact with the roads the right way as per the manufacturer specifications. In other words, it's the adjustment of the car's suspension – the system that connects the vehicle to its wheels. It's not an adjustment of the tyres or wheels themselves. It's ensuring that the wheels are 'inline' with given specifications. There are a number of adjustments in this such as toe-in, toe-out, camber and castor; which varies from vehicle to vehicle.

How to Know if your car needs Wheel Alignment?

As mentioned in the car owner's manual, wheel alignment needs to be checked at regular intervals of 5000kms to 10000kms even if one doesn't feel the need for such a service.

But again, it also depends upon the roads that the vehicles drive-in. If the road condition is rough or has lots of potholes, they need to be checked early. Even a single pothole is sufficient to throw a car out of alignment.

Thus, let us look into the signals which tell us to go for wheel alignment of our car;

- The car is pulling to one side of the road
- The tyres are squealing [long shrill sound]
- The tire tread is wearing out more on one side or unevenly
- The tire tread is wearing out prematurely
- There is unusual vibration while driving
- The steering wheel tilts off centre while you are driving
- The steering wheel vibrates when accelerating
- Else, if you have driven through a very rough patch of road and are not feeling comfortable driving as you used to

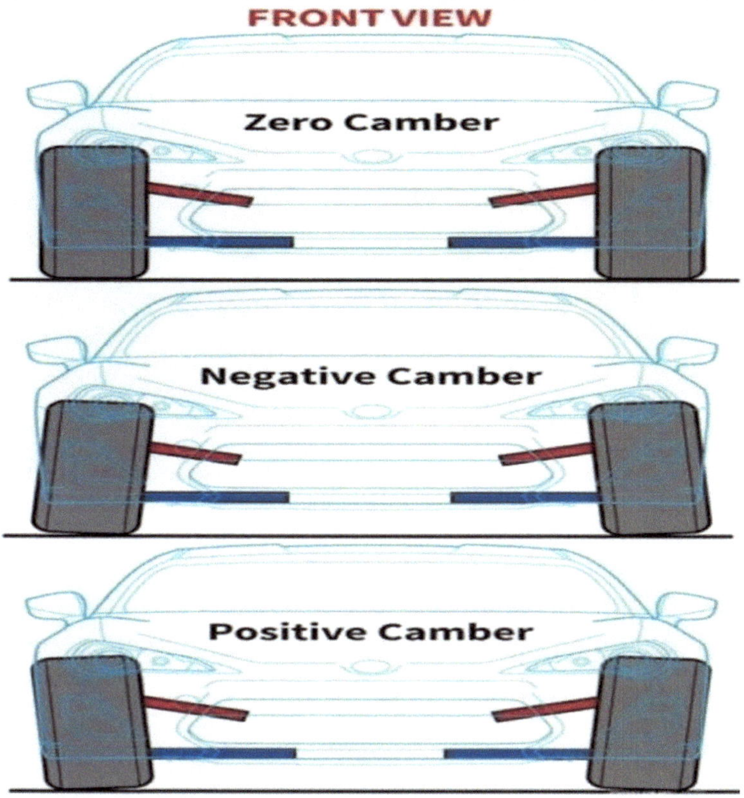

Benefits of Wheel Alignment:

- Smoother drive
- Longer tyre life
- Uniform tyre tread wear
- Optimum fuel mileage
- No vibrations or unnecessary noises
- Steering wheel in the centre

- And overall, peace of mind

4. Wheel Balancing:

A lot of car owners are still unaware of the difference between wheel alignment and wheel balancing, whether both are regularly required and the individual benefit of both.

In fact, wheel balancing is basically balancing the total wheel as a whole by fixing weights on the opposite side of the heavier portion of the rim as guided by the wheel balancer machine.

During a tyre balancing service, your wheels with tyres are mounted onto a wheel balancing machine. The machines spin the wheels to measure the imbalance so that the specialised technician can fix the correct weights to the rim to achieve a properly balanced wheel assembly.

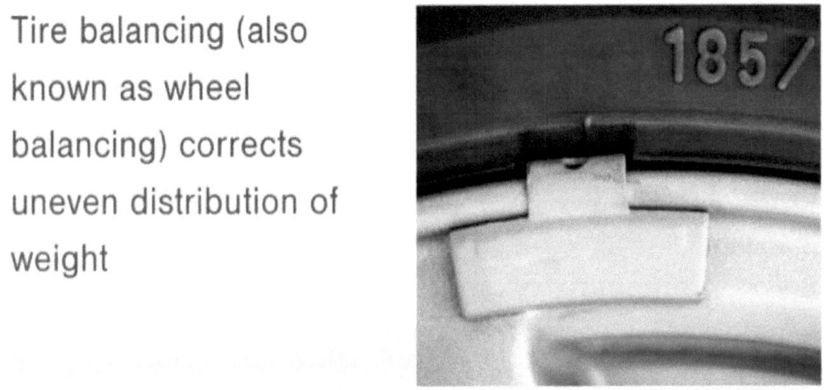

Wheel balancing corrects the uneven distribution of weights in a wheel.

Wheel balancing should be done as part of the scheduled maintenance. It's also done at the time of new tyre replacement or a flat tyre repair.

Causes of Out-of-Balance wheels:

- Uneven tyre wear
- Loss of a wheel weight when the tyre hits a curb
- A car parked for a long period
- Potholes and bumps

Signals to get Wheel Balancing done:

Even though wheel balancing should be done in scheduled intervals and along with wheel alignment and tyre rotation, still, there could be special signals which recommend us to get wheel balancing done urgently:

- When your tyres develop a cupped wear pattern
- The vibration of the vehicle at high speeds
- Uneven tread wear

Benefits of Wheel Balancing:

- Smoother drive
- Less tyre tread wear
- Uniform tyre wear
- Reduced strain on the drivetrain
- And again, peace of mind

Thus, even though, wheel alignment and wheel balancing are two different services, both are interlinked and both the services should be done at one go along with tyre rotation for optimum results. These are best left to technical experts at tyre branded sales-n-service stores but having prior knowledge helps.

5. Tyre Rotation:

Tyre rotation is the process of moving the wheels of the vehicle from one position to another on a fixed period or as and when required. Generally, it is performed while getting your car wheel alignment and wheel balancing. It is the combo of the above 3 activities that ensures uniform tyre wear and a smooth drive. Tyre rotation is done as per the manufacturer's recommendations in the car owners manual. Regularly rotating your tyres also gives you a good idea to physically inspect your tyres for any damages and to check tread depth.

Why tyre rotation?

- ➢ Even tyre wear for all tyres along with spare tyre if so intended
- ➢ To increase total tyre life
- ➢ To better fuel economy
- ➢ Smoother drive
- ➢ Uniform tread wear helps in better cornering and braking performance
- ➢ To inspect tyre tread depth and rotate accordingly
- ➢ Evenly worn tyres reduce the stress on the drivetrain
- ➢ Reduces wear and tear of expensive drive components

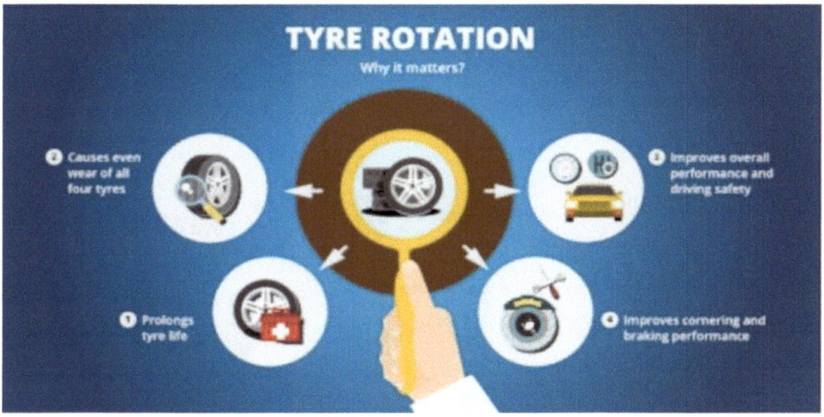

How to rotate tyres?

A common rotation pattern is to move the back wheels to the front and the front to the back but cross them when moving to the back. If the tyres are unidirectional, the tyres can only be rotated from front to back on the same side of the vehicle to preserve the rotational direction of the tyres. More complex rotational procedures are followed if the vehicle has a full-size spare tyre and if that tyre is included in the rotation.

For Front-wheel drive cars:

- Move front tyres to the rear by switching sides. The right front goes to the rear left
- Move rear wheels to the front without switching sides

For Rear-wheel drive cars:

- Move the front to rear without switching sides
- Move the rear to the front by switching sides

For Four-wheel drive cars:

- Move front tyres to the rear, switching sides
- Move rear to front and switch sides
- It's also known as the X-pattern rotation

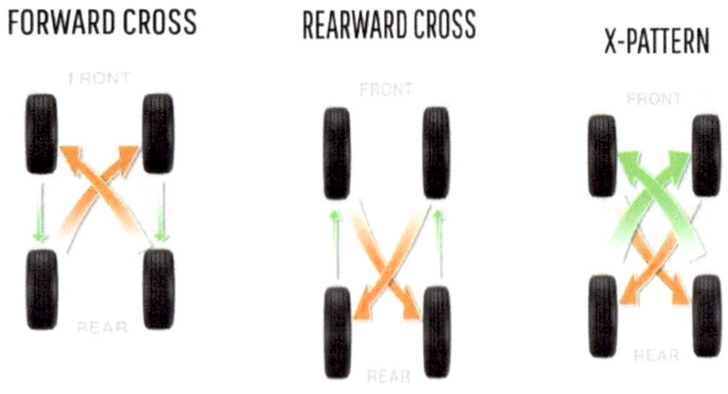

6. Good driving Habits:

   Apart from all the above 5 important tyre care tips which all can be done at once in a properly branded tyre care store, we need to follow some basic good driving habits to prolong our car's tyre life and enjoy a comfortable drive.

   - Do not drive erratically
   - Do not accelerate and apply brakes suddenly
   - Avoid hitting potholes, objects, etc
   - Slow down on bumps
   - Overtake smoothly on wide roads
   - Make turnings on ghats smoothly
   - Follow speed limits when mentioned
   - Do not drag breaks when driving downhill

7. Replace the tyre when it is needed:

   Don't drag your tyre till it will dry rot. If you do not use your vehicle very often, even if the TWI mark is not reached and you have a good amount of tread, but notice cracks developing in the sidewalls, go for new tyres replacement. It is not advisable to run on very old tyres. Normally, tyres shouldn't be used beyond 10 years.

   In other words, you shouldn't wait till your tyres are bald or till there is no tread left. Replace tyres:

   - When TWI mark is reached on the tread

- ➢ When cracks develop because of longevity
- ➢ When steel wires are exposed due to alignment issues
- ➢ When tyres have run after puncture too
- ➢ There is any sidewall cut from sharp objects
- ➢ Tyre has heavy uneven wear
- ➢ When you notice any tread separation or joint separation which may be covered under warranty if the manufacturer decides so.

8. Additional Important tyre tips:

- When installing 2 tyres, fit the tyres with the deepest tread on the rear axle.
- Never mix radial and nylon tyres on the same axle
- In the case of 2 radials and 2 nylon tyres, fit the radials on the rear axle
- Try not to install tyres with different speed ratings
- For optimum performance, go for a set of 5 tyre replacements at once and use the spare tyre too for rotation.
- Do not overload your vehicle as it puts additional pressure on tyres and will affect fuel economy too.

9. Clean car tyres periodically:

We seldom think of keeping our tyres neat and clean though we get our car washed regularly at home or hi-tech car wash centres available these days. Cleaning our tyres are important not only for the aesthetic values but also to read details on the sidewalls when required. Tyre shampoo and polish are widely available these days just as car shampoo and car detailing.

Also, it's very important to remove any small objects or stones that are trapped in the tread to avoid them from damaging the tread and the tyre in totality.

10. Important tyre accessories to have in your car:

    It is very important to have a few tyre accessories in your boot for the sake of safety and enhanced performance of your tyres. With increased awareness about maintenance, one should carry a few gadgets that would make your life hassle-free.

- Tyre Puncture repair kit:

Some of the luxury cars come with a puncture repair kit, else it would be advisable to buy one and have it in your boot as you could be travelling in a remote location where you may not find any repair shop and you will be on your own to set it right.

That's the time it will be as helpful as a godsend.

The repair kit consists of all gears to fix a small puncture like the reamer, probe, rubber glue, rubber strips, cutter, a nose-piler and maybe a few more. Depends on the kit. Once you fix your puncture, make sure to get it thoroughly checked at the nearest repair shop.

- Tyre pressure gauge:

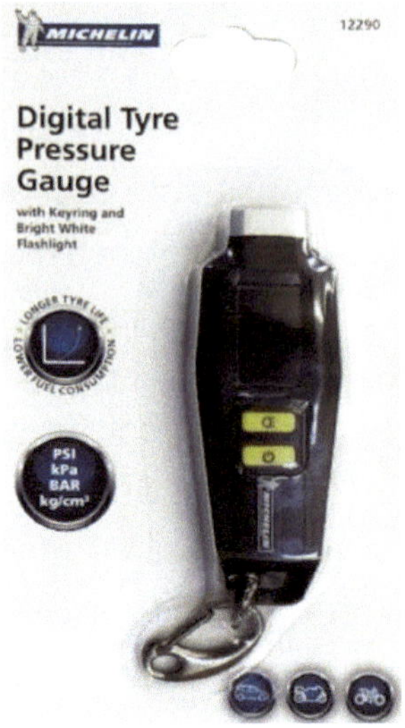

We have very well seen that maintaining optimum air pressure in the tyres results in better fuel economy, smoother drive, uniform tread wear and above everything else, it is safety. Hence always have one good quality tyre pressure gauge to check the pressure and may release them too if so necessary. A TPMS system may be installed which too will indicate any change in air pressure maybe because of a puncture or slow leakage.

- Tyre inflator:

It is again an important tool to have in your car care kit. If you have a tubeless tyre and you have suffered a puncture, a tyre inflator becomes necessary. There are a number of portable inflators available today to choose from. Many inflators come with pressure gauges too and in case of leakage of air too, this tool will come in handy.

- Valve cap:

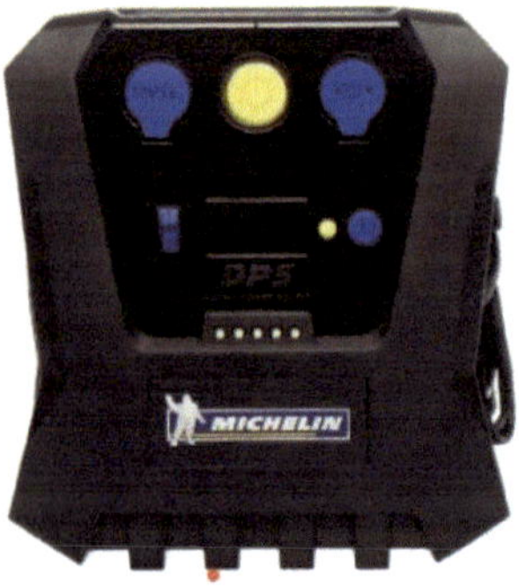

It's a small item but does a very important function of keeping the air locked inside the tyre and preventing moisture, dirt, grime from entering the tyre. Many times, due to normal tyre vibrations and jerks, the valves fall off and having extra in your kit is helpful.

- Emergency Puncture Tyre Inflator Spray: This is a pressurised liquid marketed by Michelin accessories where your deflated tyre inflates by connecting the tyre valve and the knob of the inflator.

The Definite guide in buying new tyres:

There is so much choice available today, be it in choosing a new car or just your car tyres too.

Just knowing the tyre inch size is not sufficient. This guide will help you in deciding next time you are looking to replace your car tyres with new ones.

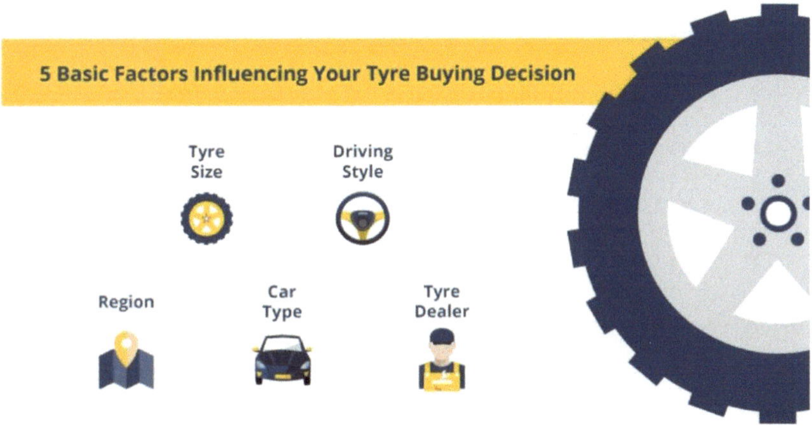

1. Tyre Size:

The tyre size and how to read a tyre size has been explained earlier which also has the highest speed limit that a tyre can reach and the maximum load a tyre can carry. The size we saw in detail was 195/55 R 16 for Hyundai i20. Tyre specifications are clearly mentioned in the sidewalls of the tyre. You can choose the tyre size as it was given by the OEM along with the vehicle or you can opt for an upsize of rim and chose wider tyre with lower aspect ratio so that the total height of the tyre more or less remains the same and the total grip of the tyre increases thereby enhancing the control of the vehicle in high speeds and turning.

2. Driving style:

The tyre comes in different designs and patterns to match different driving styles and road conditions. Driving style is how a person drives the car in general. Determining your personal driving style before you buy a new tyre is important for improved performance and a smooth driving experience. You can question yourself about:

- Do you drive at high speeds?
- Do you prefer a smooth drive?
- Do you enjoy a sporty drive?
- Do you experience stop-and-go traffic very often?
- Do you apply brakes abruptly?

Once you understand your preference, it becomes easier for you to choose tyres. For example, if you are an adventure seeker and prefer off-road journeys often, you can go for all-terrain tyres widely available in all brands or just go for high-performance tyres at high speeds if the preference is so.

3. Region:

The weather and road type in a particular region are important aspects you need to keep in mind before choosing new tyres for your car. In India, where the season keeps changing, all seasons' tyres are the best option.

But, where there are extremities in a climate like cold countries or extremely hot climates, then we need to choose tyres with grips accordingly.

Also, if the roads are quite rough, then again, we need to go for the off-roaders that provide better traction.

4. Car type:

Car type is one more factor to consider. The type of tyre you need to purchase depends on whether you are driving a plain passenger sedan, or a sporty SUV, a commercial vehicle, etc. The tread pattern, tread depth, along with tread design are set to suit different types of vehicles to provide balance, fuel efficiency, noise control, etc. You can otherwise check your owner's manual to go for the tyres that were fitted as OEM if the performance was fair enough.

5. Tyre dealer:

Today, you can visit various tyre brand stores where you can get expert guidance from their technical experts who are also going to fit your tyres into your vehicle along with the necessary services.

If you are not one of the petrolheads, it's better to leave it to them. Else, if you decided on a particular brand or pattern, you can for it.

Go for tyre change only from a recognised authorised dealer of a brand as listed in dealer locater of that brand website.

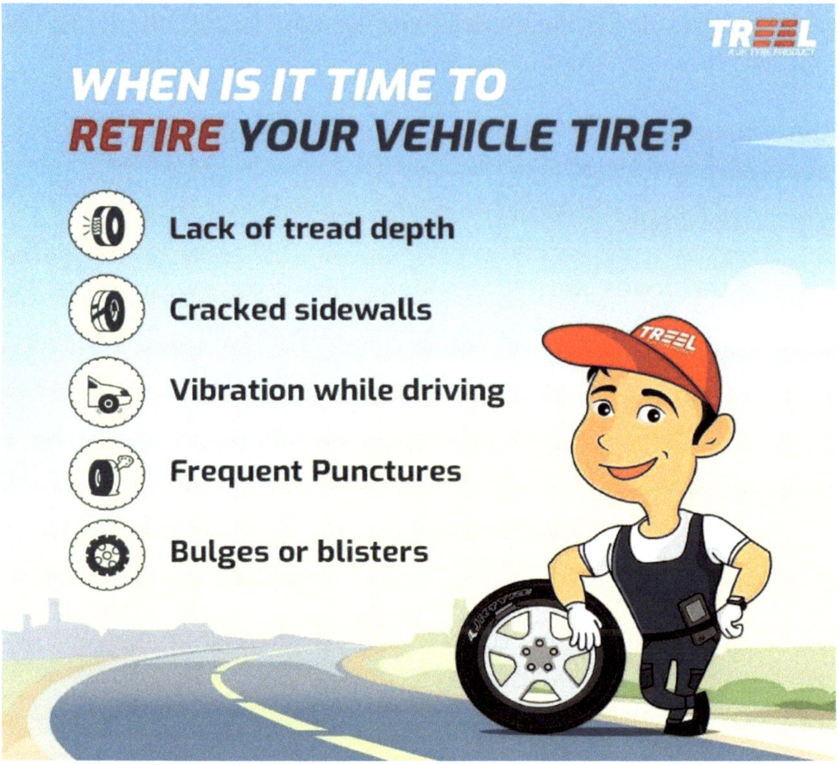

Flat tyres and their fixing procedure:

It is one of the most common problems, we all would have experienced once in our lifetime. If we are on the self-drive mode, I mean, without a chauffeur; it is in our best interest to know the basic process of changing the punctured or a flat tyre with the spare tyre. For this, we need some basic tools and a spare tyre in good condition which we should ensure before a long journey, else we can land up wasting a good amount of time. The picture below explains the process in a more informative way.

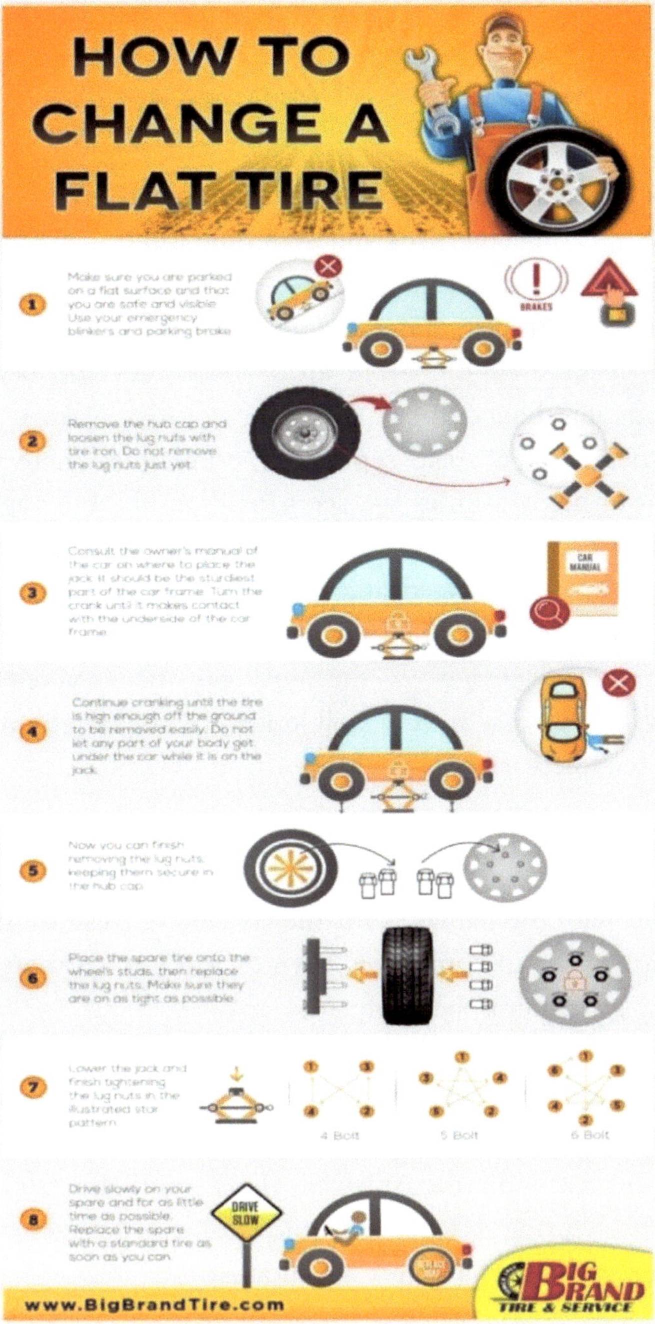
HOW TO CHANGE A FLAT TIRE
1
Make sure you are parked on a flat surface and that you are safe and visible. Use your emergency blinkers and parking brake
BRAKES
2
Remove the hub cap and loosen the lug nuts with tire iron. Do not remove the lug nuts just yet.
3
Consult the owner's manual of the car on where to place the jack. It should be the sturdiest part of the car frame. Turn the crank until it makes contact with the underside of the car frame.
CAR MANUAL
4
Continue cranking until the tire is high enough off the ground to be removed easily. Do not let any part of your body get under the car while it is on the jack.
5
Now you can finish removing the lug nuts, keeping them secure in the hub cap.
6
Place the spare tire onto the wheel's studs, then replace the lug nuts. Make sure they are on as tight as possible.
7
Lower the jack and finish tightening the lug nuts in the illustrated star pattern.
4 Bolt
5 Bolt
6 Bolt
8
Drive slowly on your spare and for as little time as possible. Replace the spare with a standard tire as soon as you can.
DRIVE SLOW
www.BigBrandTire.com
BIG BRAND
TIRE & SERVICE

CHAPTER 11

TYRE DAMAGES AND WARRANTY

We have seen all sorts of care and maintenance of a tyre but even after caring, the tires still gets punctured; sometimes we run with the punctured tyre for a long distance without knowing; sometimes there is a deep impact cut; sometimes there is leakage in the valve area; sometimes the beads are damaged and hence there might be a slow leakage of air; sometimes there is tread separation, sometimes there is joint separation; many-a-times we see a balloon-like bulging in the sidewall of the tyre. Thus, there are a host of reasons for which the tyre can get damaged and only after proper inspection at the tyre care centre, we can know the real reason for the damage. The damage to the tyre can be covered under warranty or not can be very well judged by the tyre brand store trained technicians.

The warranty policy of the company is very clear which says that in case of manufacturing defects, they will honour the customer with the damages for the remaining tread on pro rata basis. Anyways, all companies have a clear-cut warranty policy and they go as per policy only.

Major types of Tyre Damages:

1. Bulging in Sidewall: A small balloon-like formation is seen in this case on the sidewall. This could be because of manufacturing defects or could be inside impact damage, generally known as 'air' formation.
2. Puncture Running: In the case of a tube-type tyre that has run for some distance after the tyre getting punctured,

there is a circular cut along with the rim pressure points and the tyre needs to be replaced with a new one. It doesn't come under warranty. Thus, most cars run on tubeless tyres today which have more resistance power in case of a puncture.

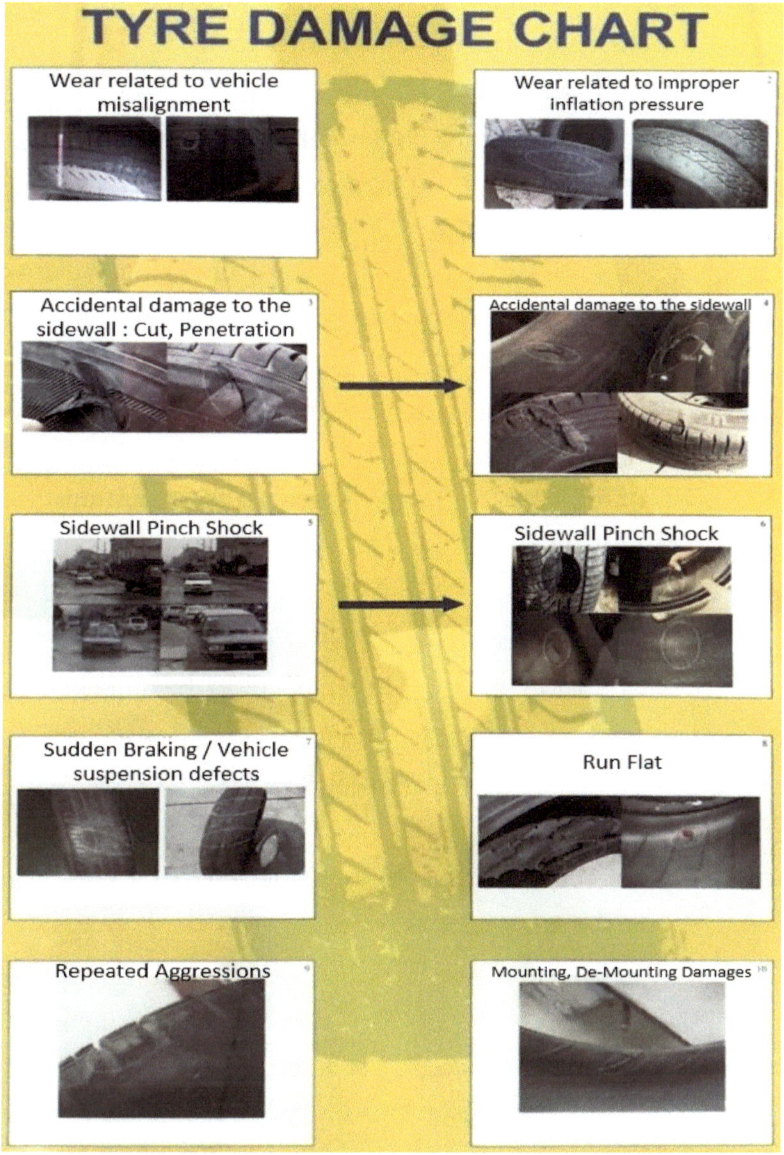

3. See through Cut in Sidewall: This is common tyre damage of radial tyres. The radial tyres, because of their construction pattern are flexible and softer at the sidewall area and hence the vehicle owner often finds large cuts or see-through cuts on the sidewall. Only minor cuts are patched up and repaired and it's better to scrap the tyre in case of bigger, see-through ones for passenger cars as safety comes first.
4. Crown area damage: It's the head of the tyre without tread and damages to this area is generally caused by bigger nails or sharper stones and other objects that the tyre hits at high speed. Generally, the crown area is below the tread and it's not easily repairable in case of larger damages and the tyres are scrapped off. In the case of larger commercial tyres, owners go for retread after repair.
5. Tread damages: Tread is the main rubber surface that is in direct contact with the road and the tyre provides traction because of this tread. Thus, it bears the maximum wear and tear damages directly. The damages in tread could be of various types:

    - One sided wears: it's because of faulty wheel alignment or suspension system. Both need to be checked and corrected. If the steel wires are exposed after the total tread is worn out, a replacement of the tyre is suggested.
    - Cupped wear out: this happens because of a fault in the balanced wheel and hence wheel balancing is recommended at the earliest.
    - Centre tread wear out: this is an inflation issue and the air pressure needs to be reduced as per recommendations and monitored.

- Both sides wear out: sometimes because of underinflation, there is wear out on the sides and shoulder area and again, air pressure needs to be monitored properly.
- Skidding damages: Some drivers take fun in skidding the vehicle but this case flattens the tread in a particular tread area unevenly and leaves the tyre unfit for use.
- Sudden Braking: tread can also get damaged if excessive sudden brakes are applied as this leads to very uneven worn out, may also lead to tyre burn or burst.

IF YOUR FRONT TIRES LOOK LIKE THIS					
TIRE PATTERN	FEATHERED EDGES	CUPPED (CUPPING)	CUPPED EDGES	EXCESSIVE WEAR/ONE SIDE	DIAGONAL WIPE
COMMON CAUSE	Improper toe	Improper balance or worn shocks	Improper balance; worn shocks with incorrect toe or camber	Incorrect camber; worn ball joints, wheel bearings or springs	Incorrect rear toe; worn bushings, grommets or ball joints

6. Bead Damages: it's the portion that holds the Rim of the wheel and hence this expands a lot to fit the tyre in the wheel. In case of manual fitment at roadside shops with inappropriate tools, the bead may get damaged and

become unfit for use. Thus, fitment should be done at proper tyre centres with fitment machines only.

7. Tread or Joint separation: it rarely happens but it does and this most probably happens because of some defect in tyre manufacturing and is generally awarded to the customer under warranty.

8. Air leakage at valves: It's better to use valve caps to avoid any leakage of air from the area.

Though, the above cover major tyre damages and unless manufacturing fault, which is generally decided by the manufacturer as per its laid down warranty policy, the owner of the vehicle has to bear the brunt of tyre damages. Hence it is always recommended to drive with intelligence and some amount of caution.

For detailed information on damages and warranty, please refer to the tyre manufacturer's warranty policies.

CHAPTER 12

TYRE RECYCLING

It is the process of recycling waste tyres that are no longer suitable to be used on vehicles due to wear or irreparable damage. These tyres are a challenging source of waste, due to the large volume produced, the durability of the tyres and the components of the tyre that are ecologically problematic. Because tyres are highly durable and non-biodegradable, they can consume valued space in landfills.

Every year, 800 million of scrap rubber tyres are disposed of worldwide, as a consequence of the huge increase in the number of vehicles on the road. In the United States alone, about 300 million scrap tyres are generated annually and India too produces 100 million scrap tyres annually. The number is expected to increase by 2% every year. The problem would become more severe due to the rapid growth in domestic consumption and a number of developing countries like China, Indonesia and so forth. It is, however, widely agreed that landfilling is not a favourable option considering land wastage and adverse environmental impacts. In light of these drawbacks, regulations have been laid down in many countries to prevent further dumping of scrap tyres. It is essential to adopt an appropriate waste management strategy to cope up with the increasing amount of scrap tyres.

Major ways of utilising scrap tyres are retreading, recycling, (in civil engineering and ground rubber application) and energy recovery (through combustion in cement kilns and pyrolysis).

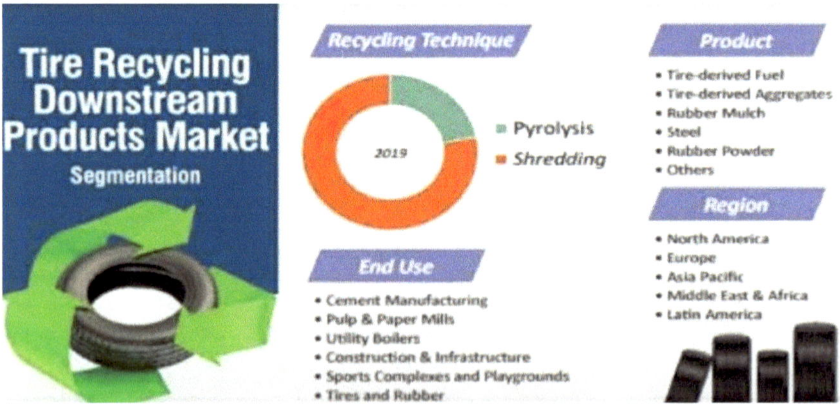

Pyrolysis and devulcanization could facilitate recycling. Aside from being used as fuel, the main end use for tyres remains ground rubber. Pyrolysis is a thermal decomposition of materials at elevated temperatures in an inert atmosphere.

Extreme pyrolysis leaves mostly carbon as a residue. In the pyrolysis process, the scrap tyres are dumped into a shredder machine which shreds the waste tyres into small pieces which are further made into almost powdery substances to be used for the ground purpose, be it in the gymnasiums, parks, pathways, athletic tracks and others.

Extracting oil from scrap tyre:

Recycling old tyres from the pyrolysis process is a beneficial option. Pyrolysis of old tyres consume waste and there is no emission of pollutants. It can also be used to recover carbon black and oil. The characteristics of fuel oil generated are similar to the oil used for diesel engines which are being tested on different capacities of diesel vehicles.

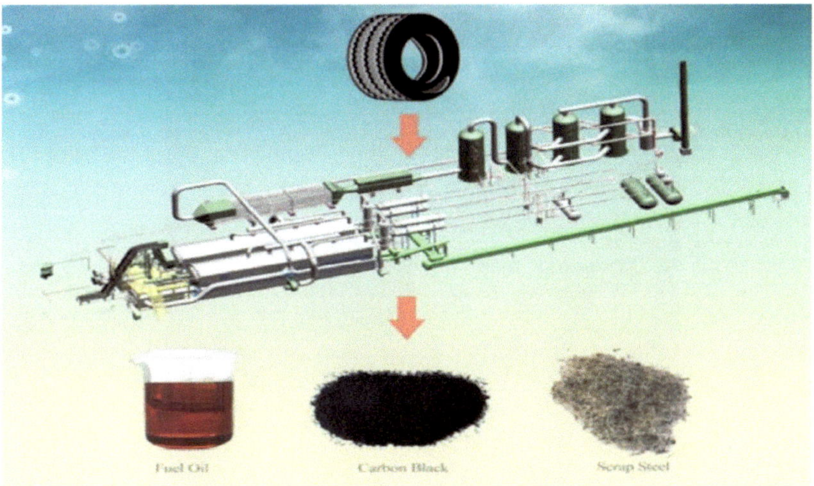

The tyre pyrolysis oil has a high calorific value of about 43Mj/Kg and can be burned directly as fuel or as supplementary feeding for refineries.

Green Distillation Technologies extracts oil by a special process where as much as 4300litres of oil can be produced from a 7000kg giant tyre. From the oil, Professor Brown produced diesel oil and tested it in Hyundai I load and found the pickup to be good.

Know more from: www.qut.edu.au

All the sources that produce waste tyres should compulsorily hand over the tyres to the nearest tyre recycling plants so that it is not burnt in the open as the gases produced are quite hazardous for the environment. Here is an example of environmentally friendly use of waste tyres.

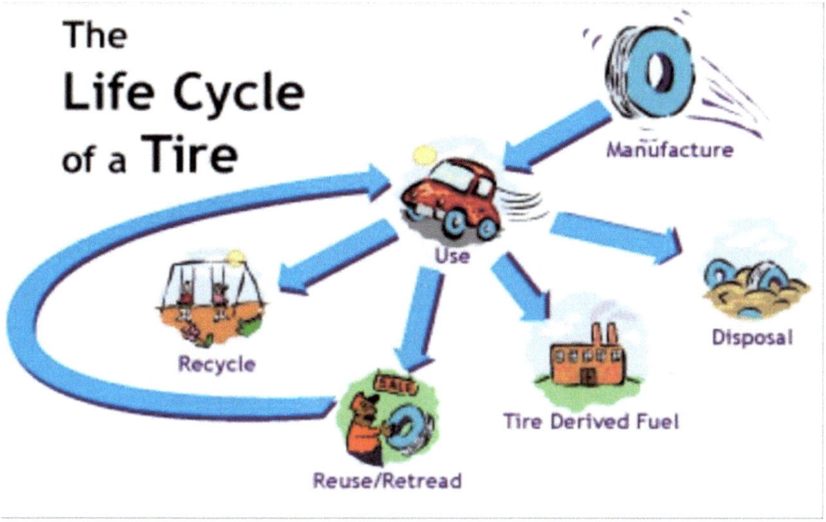

A new tyre life on the road lasts for approximately 50000kms or could be more or less depending upon its care and its quality. After the tyre tread signals for its removal, it can be given over for retreading, where a new tread is fixed on the tyre or are disposed to be used by somebody for burning which is hazardous, or could be used as a fuel in different industries, especially in cement or could be recycled for multiple uses as seen earlier. This is how the life cycle of a tyre goes. Generally, a commercial tyre is re-treaded but not passenger car tyres for safety reasons.

CHAPTER 13

TYRE RETREADING

It is the process where a new tread is fixed on the used tyre after careful inspection. Thus, in this process, you basically make your old tyres, new. This is done at proper tyre retreading factories. People generally opt for tyre retreading for their commercial vehicles or agriculture purpose vehicles as the cost of tyre retreading when compared to the cost of new tyres is less than half. It is better to avoid this process for high-performance cars as the comfort and safety levels will obviously not be as good as the new tyres. Tyres without any damages can only be re-treaded.

Types of tyre retreading:

1. Cold process:

   Although named cold process, this process makes use of heat. The tyres are heated above 100 degrees Celsius and no moulds are used and hence it is costly but still half the price of a new one.

2. Hot retreading:

   In this process, as the name suggests, tyres are heated more than 150 degrees Celsius and hence tyre shape can also be changed to some extent.

   Process of tyre retreading:

   We have seen a new tyre manufacturing process where more than 200 ingredients go into making it. But here it is not so and you are required to just add treads to the worn-out tyres.

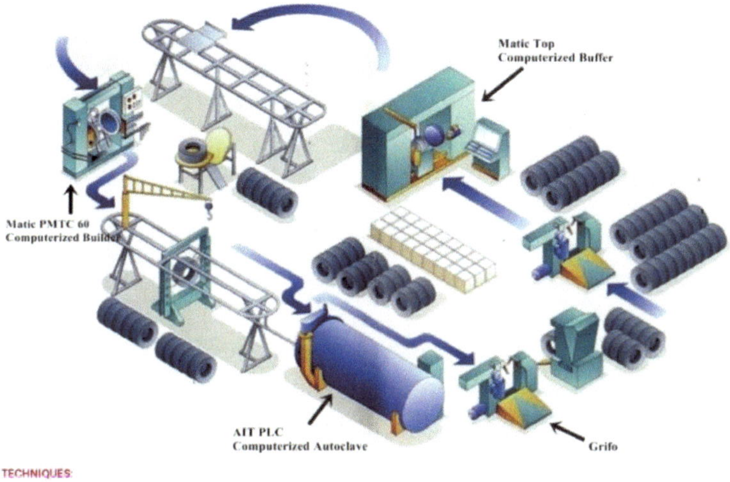

1. Tyre Inspection:

   First as soon as a tyre is received by the plant, the tyre inspector inspects manually and then if he finds it ok it goes through a test called Laser Digital Stereography which checks tyre in detail.

2. Tyre Buffing:

   Here, any leftover treads are buffed and the tread area is smoothened for the new treads to be fixed by the buffer machine.

3. Tyre Repairing:

   In this process, any minor repairs are looked after so that the problems don't arise again.

4. Tyre Building:

   In this process, with the help of computers, tyre treads are fixed accurately with the specific size as judged by the computers.

5. Curing:

   In this, the tyres are subjected to heat to check the quality of rubber.

6. Final Inspection:

   Here, retreads tyres are finally inspected if they can hit the road or not.

   Retreading of tyres are way cheaper and advisable for commercial and Agriculture vehicles. In India, nearly 70% of the MHCV (Medium and Heavy Commercial Vehicles) tyres are re-treaded, at least once before being disposed of. Some of the vehicle owners go for even 3 retreads if the casing holds the strength.

   It is not very advisable for the personal vehicle because of safety reasons. Though 2-3% of people still retread their car tyres if they intend to sell them off soon or if they have bought a pre-owned car.

CHAPTER 14

THE DIGITAL MEDIA MARKETING

All the tyre companies in India have various marketing and logistics offices spread throughout the country and in other countries too for widespread distribution to maximise reach of their products. The investment in manpower is a huge cost. Logistics or distribution offices are necessary but companies are cutting in on manpower costs with the rise of the importance of social media marketing.

All the companies very well have their own corporate websites along with country-specific websites which contain all information – mission and vision; history; products; along with proper search and find options; technical information; career information and most importantly, their dealer and brand store network so that a customer can find, call and visit the nearest one for a quick purchase which is the ultimate aim of any company – enhancing sales.

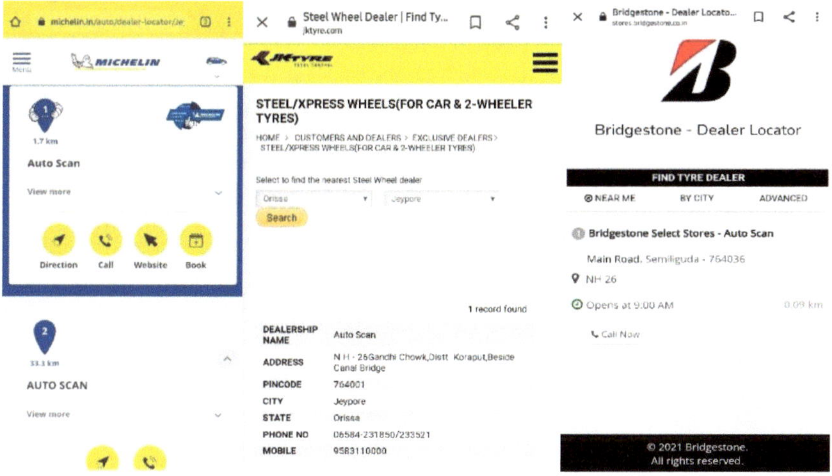

In the above pics, we can notice the dealer locator of the nearest Michelin, JKTyre and Bridgestone dealer in Koraput district. The customer can reach the dealer via the google location map even in case of lack of call connectivity.

Apart from own websites and most widely used google search, these days, other important social media marketing includes – YouTube, Facebook, LinkedIn, Instagram, Twitter among a few others.

They keep on updating the above media on a regular basis to keep their followers busy with the latest offerings from their company and also plan out various schemes to increase their follower base thereby increasing their reach.

Online tyre sales are also gaining momentum and many neutral trading portals like TyreDekho, TyreWale, are selling tyres online. Amazon, as in other items, is catching on in tyre sales too. It is because customers are dependent on value-added services like automated tyre fitment, nitrogen air, balancing and wheel alignment that they are not very keen on online purchase but from branded stores of their choice or from a multi-brand dealer where they get maximum brands and varieties to choose from.

Thus, all those products which have to be backed up by service will give online traders some sort of challenge.

Again, the transportation cost of bulky products like tyres is a challenge too. Thus, home delivery of tyres from online portals is not a very good idea as of now.

Being active and alive on the various platforms of social media has become very necessary for tyre companies or for any business to generate maximum visibility and thereby sales. On the personal front too, one can be active via YouTube or maybe via Instagram to influence society, once a good number of followers clock in.

ADVERTISEMENT & MARKETING:

Tyre advertisement has been existing ever since the birth of cycle tyres in the late 1800s. Here we can see some retro advertisements which used to feature in newspapers and as wall paintings. Other mediums like trains and vehicles too were used to advertise. In the modern-day, there are numerous other mediums too like the digital media that we saw earlier, television too is a popular audio-visual and powerful medium and we have seen brand stores of almost all brands so as to attract and offer customers the best buying experience.

Visibility in the market is the key along with good product performance to ensure repeat buying by tyre users. Let's understand why marketing is a must:

- Product visibility and repeat demand
- New product launches
- Brand value enhancement
- Pushing low selling products with schemes
- Maintain product segment leadership
- Target specific customer profiles like children

Random advertisement pictures of some top tyre brands:

The Indian tyre brands advertise via cricketers as it's a very popular sport in the country and gets a lot of visibility. Some brands focus on a particular product feature like MRF Wanderer and JK's all-terrain Ranger series whose focus is on the off-road driving capability to Ceat's focus on braking.

CHAPTER 15

FREQUENTLY ASKED QUESTION

We have tried to cover most of the topics regarding tyres, their maintenance, their damages and repair. However, let's look at a few frequently asked queries.

1. Why are the tyres black?

Carbon powder is added to milky white natural rubber to make it black for the durability and longevity of tyres. Else, white tyres used to run only 10k km earlier which now clocks 50k km.

2. Can we know the manufacturing date of the tyres?

Yes, the sidewall contains, a 4-digit mark which denotes the week and year of manufacture. For example, 2020 denotes the 20th week of the year 2020.

Generally, the shelf life of a tyre is 5 years and it is not recommended to buy tyres if it is older than 2 years.

3. Can we measure the tread depth of a tyre?

The tread depth of a tyre can be measured by a tyre tread depth gauge. A new car tyre generally has an 8mm tread that needs to be removed when it reaches the tyre wear indicator mark which is 1/16th of the total tread depth.

4. Why do tyre cracks develop without any damage?

Even if the tyres do not run, they develop cracks on the sidewalls and tread area too with the passage of time and need to be replaced if such a condition occurs. This generally occurs after 5 years from the date of manufacturing.

5. What is Aquaplaning or Hydroplaning?

This occurs when not much tread is left on the tyres and the tyre loses grip on a shallow body of water and starts to float and skids due to lack of grip. It generally happens at high speeds. That's the reason, mentioning of wet grip on sidewalls of the tyre is soon becoming mandatory.

6. Does using a wider upsize tyre reduces fuel economy?

It seems so to a common man as friction is expected to increase because of the wider tread area. But it's not so in reality. Friction or traction depends on the grooving pattern of tread and a wider tread tyre might actually lead to better fuel mileage if the overall friction reduces.

7. How is tyre storage important?

Tyres in tyre godowns should be stored in standing positions in racks or against a wall but not much in one above another in sleeping position also called tyre stacks. Tyres stored correctly in godowns well protected from extreme heat and cold will have a longer shelf life.

CHAPTER 16

THE ROAD AHEAD

The top tyre companies across the globe have to look for sustainable development and mobility keeping in mind the environmental factors, societal factors along with developing the most advanced and innovative tyre designs.

Michelin, for instance, has been testing its first airless tyre concept in collaboration with General Motors called the UPTIS [Unique Puncture-Proof Tyre System] in an airless mobility solution for passenger vehicles. This is expected to:

1. Make driving puncture-free and hence will save a considerable amount of time
2. Reduce the chance of tyre burst to almost zero owing to its utmost flexibility or elasticity feature
3. Its revolutionary design and material will yield sustainable benefits, such as a number of tyres that are left as scrap, overall raw material and energy used in tyre production.

SERIES / CONCEPT

MICHELIN UPTIS AIRLESS TIRE

Short for Unique Puncture-Proof Tire

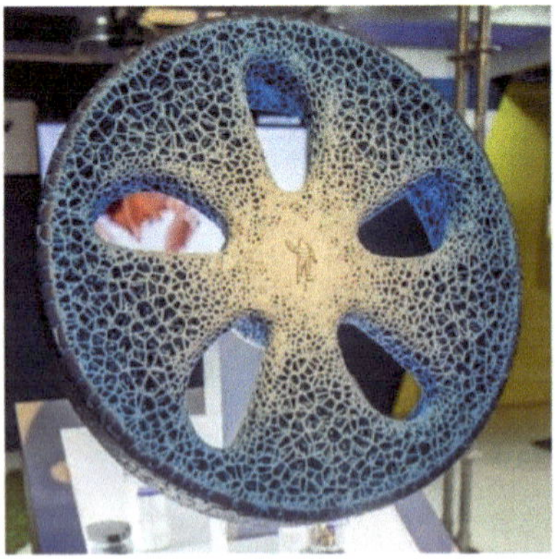

Vision concept tyre by Michelin

It is expected to be available for the replacement market by 2024 as it is still in the testing phase and mass production is subject to begin soon. This unique airless tyre is very much visible on social media for some time now. Hope it becomes a huge success for the betterment of mankind and it shall change the way we travel and lead the way towards a better and sustainable world.

Tyres & Sustainability:

When Michelin introduced radial tyres in 1946, it reduced the rolling resistance by 30%, which we still drive on today. In 1992, the addition of silica further reduced rolling resistance by 20% thereby reducing carbon emission. Thus, the raw materials that are being used today to make a tyre are very much recycled after its usage for a better, cleaner and sustainable world. An airless tyre would be a major step in this direction.

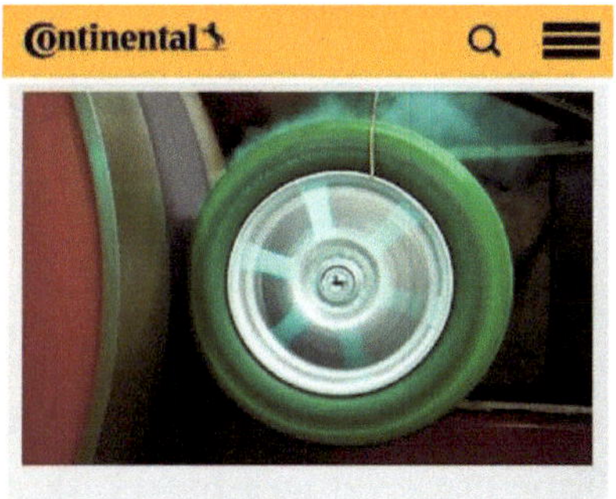

Tyres for hybrid and electric vehicles help reduce CO2 emissions.

Air-free bicycle tyre by Bridgestone:

Air-free bicycle tyre by Bridgestone

In this concept, as above, the Japanese giant proposes to use a structure of spokes that stretch the inner sides of tyres to support the user's weight. These will be made from thermoplastic resin which will be fully recyclable and will leave no residue of carbon as all tyres do. Hence, it again is a future concept expected to come into reality soon. They are developing air-free tyres for cars and testing them too just as Michelin is doing and are expected to bring them to the replacement market in the coming years.

Eagle 360 urbane tyre by Goodyear:

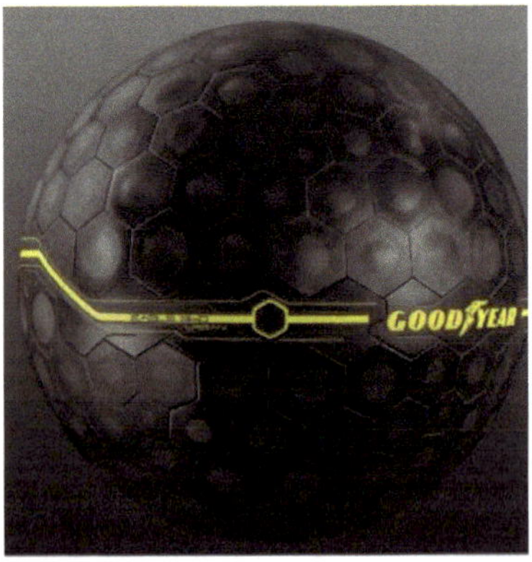

Eagle 360 Urban tyre by Goodyear

At the 2019 Geneva Motor show, Goodyear presented a spherical concept tyre for driverless cars that is designed to become the vehicle's 'nervous system'.

Driven by artificial intelligence, it will be covered in a bionic skin of sensors made from super elastic polymers which will enable it to sense driving conditions including terrain and weather.

This is a very futuristic concept and will be a great boost to the way people would travel in future.

At present too, there are a number of countries that have accepted driverless cars and are using them both on the gasoline front and on the Electric cars as well.

Electric vehicles, the future of automobiles has created a new challenge for tyres too. For Tesla, Goodyear added foam to the tyres to make them quieter as electric vehicles themselves are very quiet.

Super elastic tyre by NASA:

NASA'S non-pneumatic super elastic was initially developed for future missions to Mars but can be seen as an alternative to our ordinary air-filled tyres. It is composed of shape memory alloys that can withstand the high strain and excessive deformation without permanent damages. It has been developed by NASA in collaboration with Goodyear.

Superelastic Tyre by NASA

The new tyre norms will help make tyres safer, while allowing customers to make an informed choice

Star Ratings for Tyres too:

The government of India has proposed mandatory norms for tyre star ratings from October 2021 onwards whereby the tyres need to mention the rolling resistance, wet grip and rolling noise. Similar norms are already existing in the developed markets from 2016 where Indian tyre makers supply and hence following these safety norms should not be any hindrance to the Indian tyre manufacturers. Ceat has already started star ratings in its 'Secura' range of tyres. Existing tyre models need to comply soon as per the policies and timeframe laid down by the government.

We are living in an era where technology rules the world. Bridgestone new concept tyre is developed in the face of new futuristic vehicles, referred to as CASE – Connected, Autonomous, Shared and Electric vehicles.

These vehicles are causing a stir in the tyre manufacturers as well who are in a race to introduce Smart tyres which will integrate with CASE vehicles.

These smart tyres will be able to send real-time data to the car owners to change their driving style to maintain air pressure and temperature accordingly.

JKTyre, a major Indian tyre manufacturer are riding high on their latest Smart tyre which are coming with sensors that send real-time data to maintain correct air pressure and temperature thereby improving vehicle safety and fuel mileage to the owners mobile installed application. Launched in September 2021, it has received good acceptance in the market as the new breed of

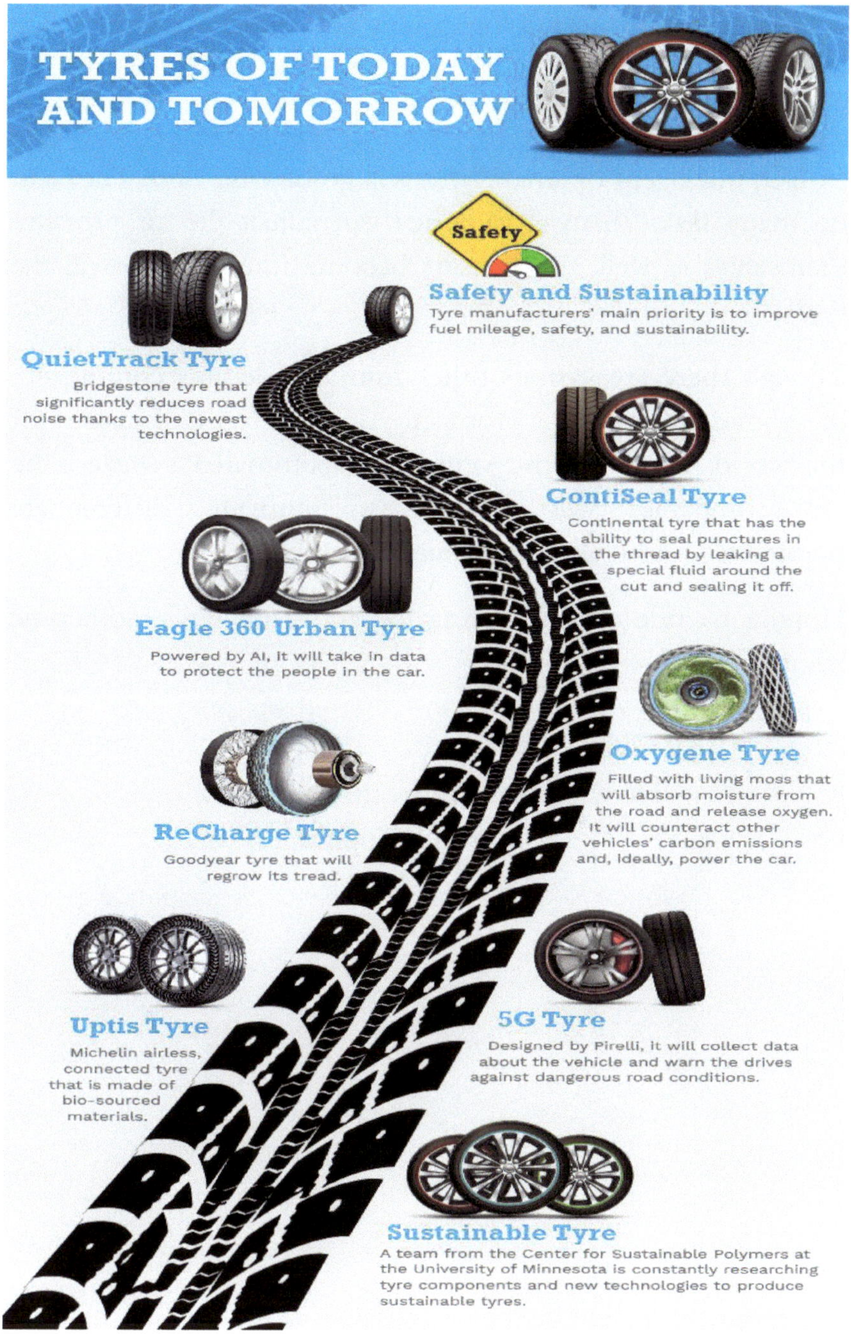

Data Credit: Whattyre.com

vehicles users are technology savvy and understand the value of being connected. This is being marketed under the theme of, "The Tyres with a Brain."

In fact, intelligent or smart tyres will go beyond alert. They can do many tasks themselves. They can adjust the air pressure themselves as well. As our cars become futuristic, so will the tyres.

Though there are various other futuristic designs coming up, the above are the most promising ones by top tyre makers of the world. From electric vehicles to automated vehicles, the tyre companies will continue to innovate, differentiate themselves for better and sustainable mobility.

Hoping to ride on these amazing futuristic tyres, sometime soon.

'He who doesn't tire, tires adversity.'
~ Martin F Tupper